The Temple of Nature

Books by Erasmus Darwin available from Timaios Press:

The Temple of Nature
The Botanic Garden
Zoonomia (2014)

Visit

www.timaiospress.com

Fact, Fiction and History of
Science and Ideas

ERASMUS DARWIN

The Temple of Nature

Or, The Origin of Society.
A Poem With Philosophical Notes

By the Author of The Botanic Garden,
of Zoonomia, and of Phytologia

Unde hominum pecudumque genus, vitaeque volantum,
Et quae marmoreo fert monstra sub aequore pontus?
Igneus est illis vigor, & caelestis origo.

[Thence are the birth of humans and animals, and the vitality of the flying / and the monsters which the ocean ferries under marbled waves / In these is an ignited vigor and origin from celestial]

Virgil: The Aeneid. Book VI, line 728

TIMAIOS PRESS

This is the unabridged text from the original edition, 1803. The author's spelling has been maintained. No changes have been done besides the following exceptions: Obvious printer's errors have been corrected. Other inconsistencies are corrected and correspond to modern practices. Most of the abbreviations of words and publication titles are here unabridged for clarity reasons. The footnotes are placed after each Canto.

Notes and additions from the editor
are in square brackets.

Editor:
Rickard Berghorn (b. 1972).
Author, literary historian, publisher, and student of history of science and ideas. See catalogues at King's Library, Stockholm, www.kb.se (Libris and Regina), and the database at Uppsala University, www.uu.diva-portal.org.

Cover Design
by Nicolas Krizan.

Timaios Press, Sweden.
www.timaiospress.com

Copyright © 2013 by Timaios Press.

All rights reserved.
Printed in Charleston, USA.

First edition from Timaios Press, November 2013.

For information about permission to reproduce selections from this book, see the Berne Convention if you live in a country which have signed the treaty. Otherwise, write to <info@timaiospress.com>.

ISBN 978-91-87611-00-1

The Temple of Nature
TABLE OF CONTENTS

PREFACE ... 1

CANTO I.
PRODUCTION OF LIFE ... 2
[Arabic numbers indicate the lines.] • I: Life, Love, and Sympathy 1. Four past Ages, a fifth beginning 9. Invocation to Love 15 • II: Bowers of Eden, Adam and Eve 33. Temple of Nature 65. Time chained by Sculpture 75. Proteus bound by Menelaus 83. Bowers of Pleasure 89. School of Venus 97. Court of Pain 105. Den of Oblivion 113. Muse of Melancholy 121. Cave of Trophonius 125. Shrine of Nature 129. Eleusinian Mysteries 137 • III: Morning 155. Procession of Virgins 159. Address to the Priestess 167. Descent of Orpheus into Hell 185 • IV: Urania 205. God the First Cause 223. Life began beneath the Sea 233. Repulsion, Attraction, Contraction, Life 235. Spontaneous Production of Minute Animals 247. Irritation, Appetency 251. Life enlarges the Earth 265. Sensation, Volition, Association 269. Scene in the Microscope; Mucor, Monas, Vibrio, Vorticella, Proteus, Mite 281 • V: Vegetables and Animals improve by Reproduction 295. Have all arisen from Microscopic Animalcules 303. Rocks of Shell and Coral 315. Islands and Continents raised by Earthquakes 321. Emigration of Animals from the Sea 327. Trapa 335. Tadpole, Musquito 343. Diodon, Lizard, Beaver, Lamprey, Remora, Whale 351. Venus rising from the Sea, emblem of Organic Nature 371. All animals are first Aquatic 385. Fetus in

the Womb 389. Animals from the Mud of the Nile 401. The Hierophant and Muse 421-450.

NOTES TO CANTO I ... 16

CANTO II.
REPRODUCTION OF LIFE ... 29

[Arabic numbers indicate the lines.] • I: Brevity of Life 1. Reproduction 13. Animals improve 31. Life and Death alternate 37. Adonis emblem of Mortal Life 45 • II: Solitary reproduction 61. Buds, Bulbs, Polypus 65. Truffle; Buds of trees how generated 71. Volvox, Polypus, Taenia, Oysters, Corals, are without Sex 83. Storge goddess of Parental Love; First chain of Society 92 • III: Female sex produced 103. Tulip bulbs, Aphis 125. Eve from Adam's rib 135 • IV: Hereditary diseases 159. Grafted trees, bulbous roots degenerate 167. Gout, Mania, Scrofula, Consumption 177. Time and Nature 185 • V: Urania and the Muse lament 205. Cupid and Psyche, the deities of sexual love 221. Speech of Hymen 239. Second chain of Society 250. Young Desire 251. Love and Beauty save the world 257. Vegetable sexes, Anthers and Stigmas salute 263. Vegetable sexual generation 271. Anthers of Vallisneria float to the Stigmas 279. Ant, Lampyris, Glow-Worm, Snail 287. Silk-Worm 293 • VI: Demon of Jealousy 307. Cocks, Quails, Stags, Boars 313. Knights of Romance 327. Helen and Paris 333. Connubial love 341. Married Birds, nests of the Linnet and Nightingale 343. Lions, Tigers, Bulls, Horses 357. Triumphal car of Cupid 361. Fish, Birds, Insects 371. Vegetables 389. March of Hymen 411. His lamp 419 • VII: Urania's advice to her Nymphs 425. Dines with the Muse on forbidden Fruit 435. Angels visit Abraham 447-458.

NOTES TO CANTO II ... 44

CANTO III.
PROGRESS OF THE MIND ... 56

[Arabic numbers indicate the lines.] • I: Urania and the Muse converse 1. Progress of the Mind 42 • II: The Four sensorial powers of Irritation,

Sensation, Volition, and Association 55. Some finer senses given to Brutes 93. And Armour 108. Finer Organ of Touch given to Man 121. Whence clear ideas of Form 125. Vision is the Language of the Touch 131. Magic Lantern 139. Surprise, Novelty, Curiosity 145. Passions, Vices 149. Philanthropy 159. Shrine of Virtue 160 • III: Ideal Beauty from the Female Bosom 163. Eros the God of Sentimental Love 177. Young Dione idolized by Eros 186. Third chain of Society 206 • IV: Ideal Beauty from curved Lines 207. Taste for the Beautiful 222. Taste for the Sublime 223. For poetic Melancholy 231. For Tragedy 241. For artless Nature 247. The Genius of Taste 259 • V: The Senses easily form and repeat ideas 269. Imitation from clear ideas 279. The Senses imitate each other 293. In dancing 295. In drawing naked Nymphs 299. In Architecture, as at St. Peter's at Rome 303. Mimickry 319 • VI: Natural Language from imitation 335. Language of Quails, Cocks, Lions, Boxers 343. Pantomime Action 357. Verbal Language from Imitation and Association 363. Symbols of ideas 371. Gigantic form of Time 385. Wings of Hermes 391 • VII: Recollection from clear ideas 395. Reason and Volition 401. Arts of the Wasp, Bee, Spider, Wren, Silk-Worm 411. Volition concerned about Means or Causes 435. Man distinguished by Language, by using Tools, labouring for Money, praying to the Deity 438. The Tree of Knowledge of Good and Evil 445 • VIII: Emotions from Imitation 461. The Seraph; Sympathy 467. Christian Morality the great bond of Society 483-496.

NOTES TO CANTO III ... 72

CANTO IV.
OF GOOD AND EVIL ... 87

[Arabic numbers indicate the lines.] • I: Few affected by Sympathy 1. Cruelty of War 11. Of brute animals, Wolf, Eagle, Lamb, Dove, Owl, Nightingale 17. Of insects, Oestrus, Ichneumon, Libellula 29. Wars of Vegetables 41. Of fish, the Shark, Crocodile, Whale 55. The World a Slaughter-house 66. Pains from Defect and from Excess of Stimulus 71. Ebriety and Superstition 77. Mania 89. Association 93. Avarice, Imposture, Ambition, Envy, Jealousy 97. Floods, Volcanoes, Earthquakes,

Famine 109. Pestilence 117. Pains from Sympathy 123 • II: Good outbalances Evil 135. Life combines inanimate Matter, and produces happiness by Irritation 145. As in viewing a Landscape 159. In hearing Music 171. By Sensation or Fancy in Dreams 183. The Patriot and the Nun 197. Howard, Moira, Burdett 205. By Volition 223. Newton, Herschel 233. Archimedes, Savery 241. Isis, Arkwright 253. Letters and Printing 265. Freedom of the Press 273. By Association 291. Ideas of Contiguity, Resemblance, and of Cause and Effect 299. Antinous 319. Cecilia 329 • III: Life soon ceases, Births and Deaths alternate 337. Acorns, Poppy-seeds, Aphises, Snails, Worms, Tadpoles, Herrings innumerable 347. So Mankind 369. All Nature teems with Life 375. Dead Organic Matter soon revives 383. Death is but a change of Form 393. Exclamation of St. Paul 403. Happiness of the World increases 405. The Phoenix 411. System of Pythagoras 417. Rocks and Mountains produced by Organic Life 429. Are Monuments of past Felicity 447. Munificence of the Deity 455 • IV: Procession of Virgins 469. Hymn to Heaven 481. Of Chaos 489. Of Celestial Love 499. Offering of Urania 517-524.

NOTES TO CANTO IV ... 104

ADDITIONAL NOTES

I.

SPONTANEOUS VITALITY
OF MICROSCOPIC ANIMALS ... 117

I: Spontaneous vital production not contrary to scripture; to be looked for only in the simplest organic beings; supposed want of analogy no argument against it, as this equally applies to all new discoveries • II: The power of reproduction distinguishes organic beings; which are gradually enlarged and improved by it • III: Microscopic animals produced from all vegetable and animal infusions; generate others like themselves by solitary reproduction; not produced from eggs; conferva fontinalis; mucor • IV: Theory of spontaneous vitality. Animal nutrition; vegetable; some organic particles have appetencies to unite, others propensities to be united; buds of trees; sexual reproduction: analogy between generation and nutrition; laws of elasticity not understood; dead animalcules recover life by heat and moisture; chaos redivivum; vorticella; shell-snails; eggs and seeds: hydra. Classes of microscopic animals; general remarks.

II.
THE FACULTIES OF THE SENSORIUM ... 127

Fibres possess a power of contraction; spirit of animation immediate cause of their contracting; stimulus of external bodies the remote cause; stimulus produces irritation; due contraction occasions pleasure; too much, or too little, pain; sensation produces desire or aversion, which constitute volition: associated motions; irritation; sensation; volition; association; sensorium.

III.
VOLCANOES ... 129

Their explosions occasioned by water falling on boiling lava; primeval earthquakes of great extent; more elastic vapours might raise islands and continents, or even throw the moon from the earth; stones falling from the sky; earthquake at, Lisbon; subterraneous fires under this island.

IV.
MUSQUITO ... 130

The larva lives chiefly in water; it may be driven away by smoke; gnats; libelulla; aestros bovis; bolts: musca chamaeleon; vomitoria.

V.
AMPHIBIOUS ANIMALS ... 132

Diodon has both lungs and gills; some amphibious quadrupeds have the foramen ovale open; perhaps it may be kept open in dogs by frequent immersion so as to render them amphibious; pearl divers; distinctions of amphibious animals; lamprey, leech; remora; whale.

VI.
HIEROGLYPHIC CHARACTERS ... 134

Used by the magi of Egypt to record discoveries in science, and historical events; astrology an early superstition; universal characters desirable; Grey's Memoria Technica; Bergeret's Botanical Nomenclature; Bishop Wilkins's Real Character and Philosophical Language.

VII.
OLD AGE AND DEATH ... 136

I: Immediate cause of the infirmities of age not yet well ascertained; must be sought in the laws of animal excitability; debility induced by inactivity of many parts of the system; organs of sense become less excitable; this ascribed to habit; may arise from deficient secretion of sensorial power; all parts of the system not changed as we advance in life • II: Means of preventing old age; warm bath; fishes; cold-blooded amphibious animals; fermented liquors injurious; also want of heat, food, and fresh air; variation of stimuli; volition; activity • III: Theory of the approach of age; surprise: novelty; why contagious diseases affect a person but once; debility; death.

VIII.
REPRODUCTION ... 146

I: Distinguishes animation from mechanism; solitary and sexual; buds and bulbs; aphises; tenia; volvox; polypus; oyster; eel; hermaphrodites • II: Sexual • III: Inferior vegetables and animals propagate by solitary generation only; next order by both; superior by sexual generation alone • IV: Animals are improved by reproduction; contagious diseases; reproduction a mystery.

IX.
STORGE ... 152

Pelicans; pigeons; instincts of animals acquired by a previous state, and transmitted by tradition; parental love originates from pleasure.

X.
EVE FROM ADAM'S RIB ... 154

Mosaic history of Paradise supposed by some to be an allegory; Egyptian philosophers, and others, supposed mankind to have been originally of both sexes united.

XI.
HEREDITARY DISEASES ... 155

Most affect the offspring of solitary reproduction: grafted trees, strawberries, potatoes; changing seed; intermarriages; hereditary diseases owing to indulgence in fermented liquors; immoderate use of common salt; improvement of progeny; hazardous to marry an heiress.

XII.
CHEMICAL THEORY OF
ELECTRICITY AND MAGNETISM ... 158

I: Attraction and repulsion • II: Two kinds of electric ether; atmospheres of electricity surround all separate bodies; atmospheres of similar kinds repel, of different kinds attract each other strongly; explode on uniting; nonconductors; imperfect conductors; perfect conductors; torpedo, gymnotus, galvanism • III: Effect of metallic points • IV: Accumulation of electric ethers by contact • V: By vicinity; Volta's electrophorus and Rennet's doubler • VI: By heat and by decomposition; the tourmalin; cats; galvanic pile; evaporation of water • VII: The spark from the conductor; electric light; not accounted for by Franklin's theory • VIII: Shock from a coated jar; perhaps an unrestrainable ethereal fluid yet unobserved; electric condensation • IX: Galvanic electricity • X: Two magnetic ethers; analogy between magnetism and electricity; differences between them • XI: Conclusion.

XIII.
ANALYSIS OF TASTE ... 189

Taste may signify the pleasures received by any of the senses, but not those which simply attend perception; four sources of pleasure in vision • I: Novelty or infrequency of visible objects; surprise • II: Repetition; beating of a drum; dancing; architecture; landscapes; picturesque; beautiful; romantic; sublime • III: Melody of colours • IV: Association of agreeable sentiments with visible objects; vision the language of touch; sentiment of beauty.

XIV.
THE THEORY AND STRUCTURE OF LANGUAGE ... 201

Ideas; words the names or symbols of ideas • I: Conjunctions and prepositions; abbreviations of other words • II: Nouns substantive • III: Adjectives, articles; participles, adverbs • IV: Verbs; progressive production of language.

XV.
ANALYSIS OF ARTICULATE SOUNDS ... 214

I: Imperfections of the present alphabet; of our orthography • II: Production of sounds • III: Structure of the alphabet; mute and antesonant consonants, and nasal liquids; sibilants and sonisibilants; orisonant liquids; four pairs of vowels; alphabet consists of thirty-one letters; speaking figure.

Preface

The Poem, which is here offered to the Public, does not pretend to instruct by deep researches of reasoning; its aim is simply to amuse by bringing distinctly to the imagination the beautiful and sublime images of the operations of Nature in the order, as the Author believes, in which the progressive course of time presented them.

The Deities of Egypt, and afterwards of Greece, and Rome, were derived from men famous in those early times, as in the ages of hunting, pasturage, and agriculture. The histories of some of their actions recorded in Scripture, or celebrated in the heathen mythology, are introduced, as the Author hopes, without impropriety into his account of those remote periods of human society.

In the Eleusinian mysteries the philosophy of the works of Nature, with the origin and progress of society, are believed to have been taught by allegoric scenery explained by the Hierophant to the initiated, which gave rise to the machinery of the following Poem.

Priory near Derby,
January 1, 1802.

Canto I.
PRODUCTION OF LIFE

I.

By firm immutable immortal laws
Impress'd on Nature by the Great First Cause,
Say, Muse! how rose from elemental strife
Organic forms, and kindled into life;
How Love and Sympathy with potent charm
Warm the cold heart, the lifted hand disarm;
Allure with pleasures, and alarm with pains,
And bind Society in golden chains.

Four past eventful Ages then recite,
And give the fifth, new-born of Time, to light;
The silken tissue of their joys disclose,
Swell with deep chords the murmur of their woes;
Their laws, their labours, and their loves proclaim,
And chant their virtues to the trump of Fame.

Immortal Love! who ere the morn of Time,
On wings outstretch'd, o'er Chaos hung sublime;
Warm'd into life the bursting egg of Night,
And gave young Nature to admiring Light!—
You! whose wide arms, in soft embraces hurl'd
Round the vast frame, connect the whirling world!
Whether immers'd in day, the Sun your throne,
You gird the planets in your silver zone;

Or warm, descending on ethereal wing,
The Earth's cold bosom with the beams of spring;
Press drop to drop, to atom atom bind,
Link sex to sex, or rivet mind to mind;
Attend my song!—With rosy lips rehearse,
And with your polish'd arrows write my verse!—
So shall my lines soft-rolling eyes engage,
And snow-white fingers turn the volant page; 30
The smiles of Beauty all my toils repay,
And youths and virgins chant the living lay.

II.

Where Eden's sacred bowers triumphant sprung,
By angels guarded, and by prophets sung,
Wav'd o'er the east in purple pride unfurl'd,
And rock'd the golden cradle of the World; *note*
Four sparkling currents lav'd with wandering tides
Their velvet avenues, and flowery sides;
On sun-bright lawns unclad the Graces stray'd,
And guiltless Cupids haunted every glade; 40
Till the fair Bride, forbidden shades among,
Heard unalarm'd the Tempter's serpent-tongue;
Eyed the sweet fruit, the mandate disobey'd,
And her fond Lord with sweeter smiles betray'd.
Conscious awhile with throbbing heart he strove,
Spread his wide arms, and barter'd life for love!—
Now rocks on rocks, in savage grandeur roll'd,
Steep above steep, the blasted plains infold;
The incumbent crags eternal tempest shrouds,
And livid light'nings cleave the lambent clouds; 50
Round the firm base loud-howling whirlwinds blow,
And sands in burning eddies dance below.

Hence ye profane!—the warring winds exclude
Unhallow'd throngs, that press with footstep rude;

But court the Muse's train with milder skies,
And call with softer voice the good and wise.
—Charm'd at her touch the opening wall divides,
And rocks of crystal form the polish'd sides;
Through the bright arch the Loves and Graces tread,
Innocuous thunders murmuring o'er their head; 60
Pair after pair, and tittering, as they pass,
View their fair features in the walls of glass;
Leave with impatient step the circling bourn,
And hear behind the closing rocks return.

Here, high in air, unconscious of the storm,
Thy temple, Nature, rears it's mystic form;
From earth to heav'n, unwrought by mortal toil,
Towers the vast fabric on the desert soil;
O'er many a league the ponderous domes extend,
And deep in earth the ribbed vaults descend; 70
A thousand jasper steps with circling sweep
Lead the slow votary up the winding steep;
Ten thousand piers, now join'd and now aloof,
Bear on their branching arms the fretted roof.

Unnumber'd ailes connect unnumber'd halls,
And sacred symbols crowd the pictur'd walls; *note*
With pencil rude forgotten days design,
And arts, or empires, live in every line.
While chain'd reluctant on the marble ground,
Indignant Time reclines, by Sculpture bound; 80
And sternly bending o'er a scroll unroll'd,
Inscribes the future with his style of gold.
—So erst, when Proteus on the briny shore, *note*
New forms assum'd of eagle, pard, or boar;
The wise Atrides bound in sea-weed thongs
The changeful god amid his scaly throngs;
Till in deep tones his opening lips at last

Reluctant told the future and the past.

Here o'er piazza'd courts, and long arcades,
The bowers of Pleasure root their waving shades;   90
Shed o'er the pansied moss a checker'd gloom,
Bend with new fruits, with flow'rs successive bloom.
Pleas'd, their light limbs on beds of roses press'd,
In slight undress recumbent Beauties rest;
On tiptoe steps surrounding Graces move,
And gay Desires expand their wings above.

Here young Dione arms her quiver'd Loves,
Schools her bright Nymphs, and practises her doves;
Calls round her laughing eyes in playful turns,
The glance that lightens, and the smile that burns;   100
Her dimpling cheeks with transient blushes dies,
Heaves her white bosom with seductive sighs;
Or moulds with rosy lips the magic words,
That bind the heart in adamantine cords.

Behind in twilight gloom with scowling mien
The demon Pain, convokes his court unseen;
Whips, fetters, flames, pourtray'd on sculptur'd stone,
In dread festoons, adorn his ebon throne;
Each side a cohort of diseases stands,
And shudd'ring Fever leads the ghastly bands;   110
O'er all Despair expands his raven wings,
And guilt-stain'd Conscience darts a thousand stings.

Deep-whelm'd beneath, in vast sepulchral caves,
Oblivion dwells amid unlabell'd graves;
The storied tomb, the laurell'd bust o'erturns,
And shakes their ashes from the mould'ring urns.—
No vernal zephyr breathes, no sunbeams cheer,

Nor song, nor simper, ever enters here;
O'er the green floor, and round the dew-damp wall,
The slimy snail, and bloated lizard crawl; 120
While on white heaps of intermingled bones
The muse of Melancholy sits and moans;
Showers her cold tears o'er Beauty's early wreck,
Spreads her pale arms, and bends her marble neck.

So in rude rocks, beside the Aegean wave,
Trophonius scoop'd his sorrow-sacred cave; *note*
Unbarr'd to pilgrim feet the brazen door,
And the sad sage returning smil'd no more.

Shrin'd in the midst majestic Nature stands,
Extends o'er earth and sea her hundred hands; 130
Tower upon tower her beamy forehead crests,
And births unnumber'd milk her hundred breasts;
Drawn round her brows a lucid veil depends,
O'er her fine waist the purfled woof descends;
Her stately limbs the gather'd folds surround,
And spread their golden selvage on the ground.

From this first altar fam'd Eleusis stole *note*
Her secret symbols and her mystic scroll;
With pious fraud in after ages rear'd
Her gorgeous temple, and the gods rever'd. 140
—First in dim pomp before the astonish'd throng,
Silence, and Night, and Chaos, stalk'd along;
Dread scenes of Death, in nodding sables dress'd,
Froze the broad eye, and thrill'd the unbreathing breast.
Then the young Spring, with winged Zephyr, leads
The queen of Beauty to the blossom'd meads;
Charm'd in her train admiring Hymen moves,
And tiptoe Graces hand in hand with Loves.
Next, while on pausing step the masked mimes

Enact the triumphs of forgotten times,   150
Conceal from vulgar throngs the mystic truth,
Or charm with Wisdom's lore the initiate youth;
Each shifting scene, some patriot hero trod,
Some sainted beauty, or some saviour god.

III.

Now rose in purple pomp the breezy dawn,
And crimson dew-drops trembled on the lawn;
Blaz'd high in air the temple's golden vanes,
And dancing shadows veer'd upon the plains.—
Long trains of virgins from the sacred grove,
Pair after pair, in bright procession move,   160
With flower-fill'd baskets round the altar throng,
Or swing their censers, as they wind along.
The fair Urania leads the blushing bands,
Presents their offerings with unsullied hands;
Pleas'd to their dazzled eyes in part unshrouds
The goddess-form;—the rest is hid in clouds.

"Priestess of Nature! while with pious awe
Thy votary bends, the mystic veil withdraw;
Charm after charm, succession bright, display,
And give the Goddess to adoring day!   170
So kneeling realms shall own the Power divine,
And heaven and earth pour incense on her shrine.

"Oh grant the Muse with pausing step to press
Each sun-bright avenue, and green recess;
Led by thy hand survey the trophied walls,
The statued galleries, and the pictur'd halls;   *note*
Scan the proud pyramid, and arch sublime,
Earth-canker'd urn, medallion green with time,
Stern busts of Gods, with helmed heroes mix'd,

And Beauty's radiant forms, that smile betwixt. 180

"Waked by thy voice, transmuted by thy wand,
Their lips shall open, and their arms expand;
The love-lost lady, and the warrior slain,
Leap from their tombs, and sigh or fight again.
—So when ill-fated Orpheus tuned to woe
His potent lyre, and sought the realms below;
Charm'd into life unreal forms respir'd,
And list'ning shades the dulcet notes admir'd.—

"Love led the Sage through Death's tremendous porch, *note*
Cheer'd with his smile, and lighted with his torch;— 190
Hell's triple Dog his playful jaws expands,
Fawns round the God, and licks his baby hands; *note*
In wondering groups the shadowy nations throng,
And sigh or simper, as he steps along;
Sad swains, and nymphs forlorn, on Lethe's brink,
Hug their past sorrows, and refuse to drink;
Night's dazzled Empress feels the golden flame
Play round her breast, and melt her frozen frame;
Charms with soft words, and sooths with amorous wiles,
Her iron-hearted Lord,—and Pluto smiles.— 200
His trembling Bride the Bard triumphant led
From the pale mansions of the astonish'd dead;
Gave the fair phantom to admiring light,—
Ah, soon again to tread irremeable night!"

IV.

Her snow-white arm, indulgent to my song,
Waves the fair Hierophant, and moves along.—
High plumes, that bending shade her amber hair,
Nod, as she steps, their silver leaves in air;
Bright chains of pearl, with golden buckles brac'd,

Clasp her white neck, and zone her slender waist;   210
Thin folds of silk in soft meanders wind
Down her fine form, and undulate behind;
The purple border, on the pavement roll'd,
Swells in the gale, and spreads its fringe of gold.

"First, if you can, celestial Guide! disclose
From what fair fountain mortal life arose,
Whence the fine nerve to move and feel assign'd,
Contractile fibre, and ethereal mind:

"How Love and Sympathy the bosom warm,
Allure with pleasure, and with pain alarm,   220
With soft affections weave the social plan,
And charm the listening Savage into Man."

"God the First cause!—in this terrene abode   *note*
Young Nature lisps, she is the child of God.   *note*
From embryon births her changeful forms improve,
Grow, as they live, and strengthen as they move.

"Ere Time began, from flaming Chaos hurl'd
Rose the bright spheres, which form the circling world;
Earths from each sun with quick explosions burst,   *note*
And second planets issued from the first.   230
Then, whilst the sea at their coeval birth,
Surge over surge, involv'd the shoreless earth;
Nurs'd by warm sun-beams in primeval caves
Organic Life began beneath the waves.

"First Heat from chemic dissolution springs,   *note*
And gives to matter its eccentric wings;
With strong Repulsion parts the exploding mass,
Melts into lymph, or kindles into gas.
Attraction next, as earth or air subsides,   *note*
The ponderous atoms from the light divides,   240

Approaching parts with quick embrace combines,
Swells into spheres, and lengthens into lines.
Last, as fine goads the gluten-threads excite,
Cords grapple cords, and webs with webs unite;
And quick Contraction with ethereal flame
Lights into life the fibre-woven frame.—
Hence without parent by spontaneous birth
Rise the first specks of animated earth;
From Nature's womb the plant or insect swims,
And buds or breathes, with microscopic limbs.

"In earth, sea, air, around, below, above,
Life's subtle woof in Nature's loom is wove;
Points glued to points a living line extends,
Touch'd by some goad approach the bending ends;
Rings join to rings, and irritated tubes
Clasp with young lips the nutrient globes or cubes;
And urged by appetencies new select,
Imbibe, retain, digest, secrete, eject.
In branching cones the living web expands,
Lymphatic ducts, and convoluted glands;
Aortal tubes propel the nascent blood,
And lengthening veins absorb the refluent flood;
Leaves, lungs, and gills, the vital ether breathe
On earth's green surface, or the waves beneath.
So Life's first powers arrest the winds and floods,
To bones convert them, or to shells, or woods;
Stretch the vast beds of argil, lime, and sand,
And from diminish'd oceans form the land!

"Next the long nerves unite their silver train,
And young Sensation permeates the brain;
Through each new sense the keen emotions dart,
Flush the young cheek, and swell the throbbing heart.
From pain and pleasure quick Volitions rise,
Lift the strong arm, or point the inquiring eyes;

With Reason's light bewilder'd Man direct,
And right and wrong with balance nice detect.
Last in thick swarms Associations spring,
Thoughts join to thoughts, to motions motions cling;
Whence in long trains of catenation flow
Imagined joy, and voluntary woe. 280

"So, view'd through crystal spheres in drops saline,
Quick-shooting salts in chemic forms combine;
Or Mucor-stems, a vegetative tribe, *note*
Spread their fine roots, the tremulous wave imbibe.
Next to our wondering eyes the focus brings
Self-moving lines, and animated rings;
First Monas moves, an unconnected point,
Plays round the drop without a limb or joint;
Then Vibrio waves, with capillary eels,
And Vorticella whirls her living wheels; 290
While insect Proteus sports with changeful form
Through the bright tide, a globe, a cube, a worm.
Last o'er the field the Mite enormous swims,
Swells his red heart, and writhes his giant limbs.

V.

"Organic Life beneath the shoreless waves *note*
Was born and nurs'd in Ocean's pearly caves;
First forms minute, unseen by spheric glass, *note*
Move on the mud, or pierce the watery mass;
These, as successive generations bloom,
New powers acquire, and larger limbs assume; 300
Whence countless groups of vegetation spring,
And breathing realms of fin, and feet, and wing.

"Thus the tall Oak, the giant of the wood,
Which bears Britannia's thunders on the flood;

The Whale, unmeasured monster of the main,
The lordly Lion, monarch of the plain,
The Eagle soaring in the realms of air,
Whose eye undazzled drinks the solar glare,
Imperious man, who rules the bestial crowd,
Of language, reason, and reflection proud, 310
With brow erect who scorns this earthy sod,
And styles himself the image of his God;
Arose from rudiments of form and sense,
An embryon point, or microscopic ens! *note*

"Now in vast shoals beneath the brineless tide, *note*
On earth's firm crust testaceous tribes reside;
Age after age expands the peopled plain,
The tenants perish, but their cells remain;
Whence coral walls and sparry hills ascend *note*
From pole to pole, and round the line extend. 320

"Next when imprison'd fires in central caves *note*
Burst the firm earth, and drank the headlong waves;
And, as new airs with dread explosion swell,
Form'd lava-isles, and continents of shell;
Pil'd rocks on rocks, on mountains mountains raised,
And high in heaven the first volcanoes blazed;
In countless swarms an insect-myriad moves *note*
From sea-fan gardens, and from coral groves;
Leaves the cold caverns of the deep, and creeps
On shelving shores, or climbs on rocky steeps. 330
As in dry air the sea-born stranger roves,
Each muscle quickens, and each sense improves;
Cold gills aquatic form respiring lungs,
And sounds aerial flow from slimy tongues.

"So Trapa rooted in pellucid tides, *note*
In countless threads her breathing leaves divides,
Waves her bright tresses in the watery mass,

And drinks with gelid gills the vital gas;
Then broader leaves in shadowy files advance,
Spread o'er the crystal flood their green expanse; 340
And, as in air the adherent dew exhales,
Court the warm sun, and breathe ethereal gales.

"So still the Tadpole cleaves the watery vale *note*
With balanc'd fins, and undulating tail;
New lungs and limbs proclaim his second birth,
Breathe the dry air, and bound upon the earth.
So from deep lakes the dread Musquito springs, *note*
Drinks the soft breeze, and dries his tender wings,
In twinkling squadrons cuts his airy way,
Dips his red trunk in blood, and man his prey. 350

"So still the Diodons, amphibious tribe, *note*
With two-fold lungs the sea or air imbibe;
Allied to fish, the lizard cleaves the flood
With one-cell'd heart, and dark frigescent blood;
Half-reasoning Beavers long-unbreathing dart
Through Erie's waves with perforated heart;
With gills and lungs respiring Lampreys steer,
Kiss the rude rocks, and suck till they adhere;
The lazy Remora's inhaling lips,
Hung on the keel, retard the struggling ships; 360
With gills pulmonic breathes the enormous Whale,
And spouts aquatic columns to the gale;
Sports on the shining wave at noontide hours, *note*
And shifting rainbows crest the rising showers.

"So erst, ere rose the science to record
In letter'd syllables the volant word;
Whence chemic arts, disclosed in pictured lines,
Liv'd to mankind by hieroglyphic signs;
And clustering stars, pourtray'd on mimic spheres,
Assumed the forms of lions, bulls, and bears; 370

—So erst, as Egypt's rude designs explain,   *note*
Rose young Dione from the shoreless main;   *note*
Type of organic Nature! source of bliss!
Emerging Beauty from the vast abyss!
Sublime on Chaos borne, the Goddess stood,
And smiled enchantment on the troubled flood;
The warring elements to peace restored,
And young Reflection wondered and adored."

Now paused the Nymph,—The Muse responsive cries,
Sweet admiration sparkling in her eyes,   380
"Drawn by your pencil, by your hand unfurl'd,
Bright shines the tablet of the dawning world;
Amazed the Sea's prolific depths I view,
And Venus rising from the waves in You!

"Still Nature's births enclosed in egg or seed
From the tall forest to the lowly weed,
Her beaux and beauties, butterflies and worms,
Rise from aquatic to aerial forms.
Thus in the womb the nascent infant laves
Its natant form in the circumfluent waves;   390
With perforated heart unbreathing swims,
Awakes and stretches all its recent limbs;   *note*
With gills placental seeks the arterial flood,   *note*
And drinks pure ether from its Mother's blood.
Erewhile the landed Stranger bursts his way,
From the warm wave emerging into day;
Feels the chill blast, and piercing light, and tries
His tender lungs, and rolls his dazzled eyes;   *note*
Gives to the passing gale his curling hair,
And steps a dry inhabitant of air.   400

"Creative Nile, as taught in ancient song,
So charm'd to life his animated throng;
O'er his wide realms the slow-subsiding flood

Left the rich treasures of organic mud;
While with quick growth young Vegetation yields
Her blushing orchards, and her waving fields;
Pomona's hand replenish'd Plenty's horn,
And Ceres laugh'd amid her seas of corn.—
Bird, beast, and reptile, spring from sudden birth,
Raise their new forms, half-animal, half-earth;   410
The roaring lion shakes his tawny mane,
His struggling limbs still rooted in the plain;
With flapping wings assurgent eagles toil
To rend their talons from the adhesive soil;
The impatient serpent lifts his crested head,
And drags his train unfinish'd from the bed.—
As Warmth and Moisture blend their magic spells,   *note*
And brood with mingling wings the slimy dells;
Contractile earths in sentient forms arrange,
And Life triumphant stays their chemic change."   420

Then hand in hand along the waving glades
The virgin Sisters pass beneath the shades;
Ascend the winding steps with pausing march,
And seek the Portico's susurrant arch;
Whose sculptur'd architrave on columns borne
Drinks the first blushes of the rising morn,
Whose fretted roof an ample shield displays,
And guards the Beauties from meridian rays.
While on light step enamour'd Zephyr springs,
And fans their glowing features with his wings,   430
Imbibes the fragrance of the vernal flowers,
And speeds with kisses sweet the dancing Hours.

Urania, leaning with unstudied grace,
Rests her white elbow on a column's base;
Awhile reflecting takes her silent stand,
Her fair cheek press'd upon her lily hand;
Then, as awaking from ideal trance,

On the smooth floor her pausing steps advance,
Waves high her arm, upturns her lucid eyes,
Marks the wide scenes of ocean, earth, and skies; 440
And leads, meandering as it rolls along
Through Nature's walks, the shining stream of Song.

First her sweet voice in plaintive accents chains
The Muse's ear with fascinating strains;
Reverts awhile to elemental strife,
The change of form, and brevity of life;
Then tells how potent Love with torch sublime
Relights the glimmering lamp, and conquers Time.
—The polish'd walls reflect her rosy smiles,
And sweet-ton'd echoes talk along the ailes. 450

End of Canto I.

NOTES TO CANTO I.

Cradle of the world ... (line 36). The nations, which possess Europe and a part of Asia and of Africa, appear to have descended from one family; and to have had their origin near the banks of the Mediterranean, as probably in Syria, the site of Paradise, according to the Mosaic history. This seems highly probable from the similarity of the structure of the languages of these nations, and from their early possession of similar religions, customs, and arts, as well as from the most ancient histories extant. The two former of these may be collected from Lord Monboddo's learned work on the Origin of Language, and from Mr. Bryant's curious account of Ancient Mythology.

The use of iron tools, of the bow and arrow, of earthen vessels to boil water in, of wheels for carriages, and the arts of cultivating wheat, of coagulating milk for cheese, and of spinning vegetable fibres for clothing, have been known in all European countries, as long as their histories have existed; besides the similarity of the texture of their languages, and of many words in them; thus the word sack is said to mean a bag in all of them, as σακκον in Greek, saccus in Latin, sacco in Italian, sac in French, and sack in English and German.

Other families of mankind, nevertheless, appear to have arisen in other parts of the habitable earth, as the language of the Chinese is said not to resemble those of this part of the world in any respect. And the inhabitants of the islands of the South-Sea had neither the use of iron tools nor of the bow, nor of wheels, nor of spinning, nor had learned to coagulate milk, or to boil water, though the domestication of fire seems to have been the first great discovery that distinguished mankind from the bestial inhabitants of the forest.

Pictur'd walls... (line 76). The application of mankind, in the early ages of society, to the imitative arts of painting, carving, statuary, and the casting of figures in metals, seems to have preceded the discovery of letters; and to have been used as a written language to convey intelligence to their distant friends, or to transmit to posterity the history of themselves, or of their discoveries. Hence the origin of the hieroglyphic figures which crowded the walls of the temples of antiquity; many of which may be seen in the tablet of Isis in the works of Montfaucon; and some of them are still used in the sciences of chemistry and astronomy, as the characters for the metals and planets, and the figures of animals on the celestial globe

So erst, when Proteus... (line 83). It seems probable that Proteus was the name of a hieroglyphic figure representing Time; whose form was perpetually changing, and who could discover the past events of the world, and predict the future. Herodotus does not doubt but that Proteus was an Egyptian king or deity; and Orpheus calls him the principle of all things, and the most ancient of the gods; and adds, that he keeps the keys of Nature, Danet's Dictionary, all which might well accord with a figure representing Time.

Trophonius scoop'd... (line 126). Plutarch mentions, that prophecies of evil events were uttered from the cave of Trophonius; but the allegorical story, that whoever entered this cavern were never again seen to smile, seems to have been designed to warn the contemplative from considering too much the dark side of nature. Thus an ancient poet is said to have written a poem on the miseries of the world, and to have thence become so unhappy as to destroy himself. When we reflect on the perpetual destruction of organic life, we should also recollect, that it is perpetually renewed in other forms by the same materials, and thus the sum total of the happiness of the world continues undiminished; and that a philosopher may thus smile again on turning his eyes from the coffins of nature to her cradles.

Fam'd Eleusis stole ... (line 137). The Eleusinian mysteries were invented in Egypt, and afterwards transferred into Greece along with most of the other early arts and religions of Europe. They seem to have consisted of scenical representations of the philosophy and religion of those times, which had previously been painted in hieroglyphic figures to perpetuate them before the discovery of letters; and are well explained in Dr. Warburton's divine legation of Moses; who believes with great probability, that Virgil in the sixth book of the Aeneid has described a part of these mysteries in his account of the Elysian fields.

In the first part of this scenery was represented Death, and the destruction of all things; as mentioned in the note on the Portland Vase in the Botanic Garden. Next the marriage of Cupid and Psyche seems to have shown the reproduction of living nature; and afterwards the procession of torches, which is said to have constituted a part of the mysteries, probably signified the return of light, and the resuscitation of all things.

Lastly, the histories of illustrious persons of the early ages seem to have been enacted; who were first represented by hieroglyphic figures, and afterwards became the gods and goddesses of Egypt, Greece, and Rome. Might not such a dignified pantomime be contrived, even in this age, as might strike the spectators with awe, and at the same time explain many philosophical truths by adapted imagery, and thus both amuse and instruct?

The statued galleries ... (line 176). The art of painting has appeared in the early state of all societies before the invention of the alphabet. Thus when the Spanish adventurers, under Cortez, invaded America, intelligence of their debarkation and movements was daily transmitted to Montezuma, by drawings, which corresponded with the Egyptian hieroglyphics. The antiquity of statuary appears from the Memnon and sphinxes of Egypt; that of casting figures in metals from the golden calf of Aaron; and that of carving in wood from the idols or household gods, which Rachel stole from her father Laban, and hid beneath her garments as she sat upon the straw. Genesis, chapter XXXI, verse 34.

Love led the Sage ... (line 189). This description is taken from the figures on the Barbarini, or Portland Vase, where Eros, or Divine Love, with his torch precedes the manes through the gates of Death, and reverting his smiling countenance invites him into the Elysian fields.

Fawns round the God ... (line 192). This idea is copied from a painting of the descent of Orpheus, by a celebrated Parisian artist.

God the first cause ... (line 223).

Ab Jove principium, musae! Jovis omnia plena.
[From Jove, ye Muses, let us begin! All Things are full of Jove.]
Virgil.

In him we live, and move, and have our being.
St. Paul.

Young Nature lisps ... (line 224). The perpetual production and increase of the strata of limestone from the shells of aquatic animals; and of all those incumbent on them from the recrements of vegetables and of terrestrial animals, are now well understood from our improved knowledge of geology; and show, that the solid parts of the globe are gradually enlarging, and consequently that it is young; as the fluid parts are not yet all converted into solid ones. Add to this, that some parts of the earth and its inhabitants appear younger than others; thus the greater height of the mountains of America seems to show that continent to be less ancient than Europe, Asia, and Africa; as their summits have been less washed away, and the wild animals of America, as the tigers and crocodiles, are said to be less perfect in respect to their size and strength; which would show them to be still in a state of infancy, or of progressive improvement. Lastly, the progress of mankind in arts and sciences, which continues slowly to extend, and to increase, seems to evince the youth of human society; whilst the unchanging state of the societies of some insects, as of the bee, wasp, and ant, which is usually ascribed to instinct, seems to evince the longer existence, and greater maturity of those societies. The juvenility of the earth shows, that it has had a beginning or birth, and is a strong natural argument evincing the existence of a cause of its production, that is of the Deity.

Earths from each sun ... (line 229). See Botanic Garden, Vol. I, Canto I, line 107.

First Heat from chemic ... (line 235). The matter of heat is an ethereal fluid, in which all things are immersed, and which constitutes the general power of repulsion; as appears in explosions which are produced by the sudden evolution of combined heat, and by the expansion of all bodies by the slower diffusion of it in its uncombined state. Without heat all the matter of the world would be condensed into a point by the power of attraction; and neither fluidity nor life could exist. There are also particular powers of repulsion, as those of magnetism and electricity, and of chemistry, such as oil and water; which last may be as numerous as the particular attractions which constitute chemical affinities; and may both of them exist as atmospheres round the individual particles of matter; see Botanic Garden, Vol. I, Additional Note VII on elementary heat.

Attraction next ... (line 239). The power of attraction may be divided into general attraction, which is called gravity; and into particular attraction, which is termed chemical affinity. As nothing can act where it does not exist, the power of gravity must be conceived as extending from the sun to the planets, occupying that immense space; and may therefore be considered as an ethereal fluid, though not cognizable by our senses like heat, light, and electricity.

Particular attraction, or chemical affinity, must likewise occupy the spaces between the particles of matter which they cause to approach each other. The power of gravity may therefore be called the general attractive ether, and the matter of heat may be called the general repulsive ether; which constitute the two great agents in the changes of inanimate matter.

And quick Contraction ... (line 245). The power of contraction, which exists in organized bodies, and distinguishes life from inanimation, appears to consist of an ethereal fluid which resides in the brain and nerves of living bodies, and is expended in the act of shortening their fibres. The attractive and repulsive ethers require only the vicinity of bodies for the exertion of their activity, but the contractive ether requires at first the contact of a goad or stimulus, which appears to draw it off from the contracting fibre, and to excite the sensorial power of irritation. These contractions of animal fibres are afterwards excited or repeated by the sensorial powers of sensation, volition, or association, as explained at large in Zoonomia, Vol. I.

There seems nothing more wonderful in the ether of contraction producing the shortening of a fibre, than in the ether of attraction causing two bodies to approach each other. The former indeed seems in some measure to resemble the latter, as it probably occasions the minute particles of the fibre to approach into absolute or adhesive contact, by withdrawing from them their repulsive atmospheres; whereas the latter seems only to cause particles of matter to approach into what is popularly called contact, like the particles of fluids; but which are only in the vicinity of each other, and still retain their repulsive atmospheres, as may be seen in riding through shallow water by the number of minute globules of it thrown up by the horses feet, which roll far on its surface; and by the difficulty with which small globules of mercury poured on the surface of a quantity of it can be made to unite with it.

Spontaneous birth ... (line 247). See Additional Note I.

In branching cones ... (line 259). The whole branch of an artery or vein may be considered as a cone, though each distinct division of it is a cylinder. It is probable that the amount of the areas of all the small branches from one trunk may equal that of the trunk, otherwise the velocity of the blood would be greater in some parts than in others, which probably only exists when a part is compressed or inflamed.

Absorb the refluent flood ... (line 262). The force of the arterial impulse appears to cease, after having propelled the blood through the capillary vessels; whence the venous circulation is owing to the extremities of the veins absorbing the blood, as those of the lymphatics absorb the fluids. The great force of absorption is well elucidated by Dr. Hales's experiment on the rise of the sap-juice in a vine-stump; see Zoonomia, Vol. I, Sect. XXIII.

And from diminish'd oceans ... (line 268). The increase of the solid parts of the globe by the recrements of organic bodies, as limestone rocks from shells and bones, and the beds of clay, marl, coals, from decomposed woods, is now well known to those who have attended to modern geology; and Dr. Halley, and others, have endeavoured to show, with great probability, that the ocean has decreased in quantity during the short time which human history has existed. Whence it appears, that the exertions of vegetable and animal life convert the fluid parts of the globe into solid ones; which is probably effected by combining the matter of heat with the other elements, instead of suffering it to remain simply diffused amongst them, which is a curious conjecture, and deserves further investigation.

Next the long nerves ... (line 269). See Additional Note II.

And young Sensation ... (line 270). Both sensation and volition consist in an affection of the central part of the sensorium, or of the whole of it; and hence cannot exist till the nerves are united in the brain. The motions of a limb of any animal cut from the body, are therefore owing to irritation, not to sensation or to volition. For the definitions of irritation, sensation, volition, and association, see Additional Note II.

Or Mucor-stems ... (line 283). Mucor or mould in its early state is properly a microscopic vegetable, and is spontaneously produced on the scum of all decomposing organic matter. The Monas is a moving speck, the Vibrio an undulating wire, the Proteus perpetually changes its shape, and the Vorticella has wheels about its mouth, with which it makes an eddy, and is supposed thus to draw into its throat invisible animalcules. These names are from Linneus and Muller; see Appendix to Additional Note I.

Beneath the shoreless waves ... (line 295). The earth was originally covered with water, as appears from some of its highest mountains, consisting of shells cemented together by a solution of part of them, as the limestone rocks of the Alps; Ferber's Travels. It must be therefore concluded, that animal life began beneath the sea.

Nor is this unanalogous to what still occurs, as all quadrupeds and mankind in their embryon state are aquatic animals; and thus may be said to resemble gnats and frogs. The fetus in the uterus has an organ called the placenta, the fine extremities of the vessels of which permeate the arteries of the uterus, and the blood of the fetus becomes thus oxygenated from the passing stream of the maternal arterial blood; exactly as is done by the gills of fish from the stream of water, which they occasion to pass through them.

But the chicken in the egg possesses a kind of aerial respiration, since the extremities of its placental vessels terminate on a membranous bag, which contains air, at the broad end of the egg; and in this the chick in the egg differs from the fetus in the womb, as there is in the egg no circulating maternal blood for the insertion of the extremities of its respiratory vessels, and in this also I suspect that the eggs of birds differ from the spawn of fish; which latter is immersed in water, and which has probably the extremities of its respiratory organ inserted into the soft membrane which covers it, and is in contact with the water.

First forms minute ... (line 297). See Additional Note I, on Spontaneous Vitality.

An embryon point ... (line 314). The arguments showing that all vegetables and animals arose from such a small beginning, as a living point or living fibre, are detailed in Zoonomia, Sect. XXXIX, 4, 8, on Generation.

Brineless tide ... (line 315). As the salt of the sea has been gradually accumulating, being washed down into it from the recrements of animal and vegetable bodies, the sea must originally have been as fresh as river water; and as it is not saturated with salt, must become annually saline. The sea-water about our island contains at this time from about one twenty-eighth to one thirtieth part of sea salt, and about one eightieth of magnesian salt; Brownrigg on Salt.

Whence coral walls ... (line 319). An account of the structure of the earth is given in Botanic Garden, Vol. I, Additional Notes, XVI, XVIII, XIX, XX, XXIII, XXIV.

Next when imprison'd fires ... (line 321). See Additional Note III.

An insect-myriad moves ... (line 327). After islands or continents were raised above the primeval ocean, great numbers of the most simple animals would attempt to seek food at the edges or shores of the new land, and might thence gradually become amphibious; as is now seen in the frog, who changes from an aquatic animal to an amphibious one; and in the gnat, which changes from a natant to a volant state.

At the same time new microscopic animalcules would immediately commence wherever there was warmth and moisture, and some organic matter, that might induce putridity. Those situated on dry land, and immersed in dry air, may gradually acquire new powers to preserve their existence; and by innumerable successive reproductions for some thousands, or perhaps millions of ages, may at length have produced many of the vegetable and animal inhabitants which now people the earth.

As innumerable shell-fish must have existed a long time beneath the ocean, before the calcareous mountains were produced and elevated; it is also probable, that many of the insect tribes, or less complicate animals, existed long before the quadrupeds or more complicate ones, which in some measure accords with the theory of Linneus in respect to the vegetable world; who thinks, that all the plants now extant arose from the conjunction and reproduction of about sixty different vegetables, from which he constitutes his natural orders.

As the blood of animals in the air becomes more oxygenated in their lungs, than that of animals in water by their gills; it becomes of a more scarlet colour, and from its greater stimulus the sensorium seems to produce quicker motions and finer sensations; and as water is a much better vehicle for vibrations or sounds than air, the fish, even when dying in pain, are mute in the atmosphere, though it is probable that in the water they may utter sounds to be heard at a considerable distance. See on this subject, Botanic Garden, Vol. I, Canto IV, line 176, Note.

So Trapa rooted ... (line 335). The lower leaves of this plant grow under water, and are divided into minute capillary ramifications; while the upper leaves are broad and round, and have air bladders in their footstalks to support them above the surface of the water. As the aerial leaves of vegetables do the office of lungs, by exposing a large surface of vessels with their contained fluids to the influence of the air; so these aquatic leaves answer a similar purpose like the gills of fish, and perhaps gain from water a similar material. As the material thus necessary to life seems to be more easily acquired from air than from water, the sub-aquatic leaves of this plant and of sisymbrium, oenanthe, ranunculus aquatilis, water crow-foot, and some others, are cut into fine divisions to increase the surface, whilst those above water are undivided; see Botanic Garden, Vol. II, Canto IV, line 204, Note.

Few of the water plants of this country are used for economical purposes, but the ranunculus fluviatilis may be worth cultivation; as on the borders of the river Avon, near Ringwood, the cottagers cut this plant every morning in boats, almost all the year round, to feed their cows, which appear in good condition, and give a due quantity of milk; see a paper from Dr. Pultney in the Transactions of the Linnean Society, Vol. V.

So still the Tadpole ... (line 343). The transformation of the tadpole from an aquatic animal into an aerial one is abundantly curious, when first it is hatched from the spawn by the warmth of the season, it resembles a fish; it afterwards puts forth legs, and resembles a lizard; and finally losing its tail, and acquiring lungs instead of gills, becomes an aerial quadruped.

The Rana temporaria of Linneus lives in the water in spring, and on the land in summer, and catches flies. Of the Rana paradoxa the larva or tadpole is as large as the frog, and dwells in Surinam, whence the mistake of Merian and of Seba, who call it a frog fish. The esculent frog is green, with three yellow lines from the mouth to the anus; the back transversely gibbous, the hinder feet palmated; its more frequent croaking in the evenings is said to foretell rain. Linneus, Systema Naturae, Art. Rana.

Linneus asserts in his introduction to the class Amphibia, that frogs are so nearly allied to lizards, lizards to serpents, and serpents to fish, that the boundaries of these orders can scarcely be ascertained.

The dread Musquito springs ... (line 347). See Additional Note IV.

So still the Diodon ... (line 351). See Additional Note V.

At noontide hours ... (line 363). The rainbows in our latitude are only seen in the mornings or evenings, when the sun is not much more than forty-two degrees high. In the more northern latitudes, where the meridian sun is not more than forty-two degrees high, they are also visible at noon.

As Egypt's rude design ... (line 371). See Additional Note VI.

Rose young Dione ... (line 372). The hieroglyphic figure of Venus rising from the sea supported on a shell by two tritons, as well as that of Hercules armed with a club, appear to be remains of the most remote antiquity. As the former is devoid of grace, and of the pictorial art of design, as one half of the group exactly resembles the other; and as that of Hercules is armed with a club, which was the first weapon.

The Venus seems to have represented the beauty of organic Nature rising from the sea, and afterwards became simply an emblem of ideal beauty; while the figure of Adonis was probably designed to represent the more abstracted idea of life or animation. Some of these hieroglyphic designs seem to evince the profound investigations in science of the Egyptian philosophers, and to have outlived all written language; and still constitute the symbols, by which painters and poets give form and animation to abstracted ideas, as to those of strength and beauty in the above instances.

Awakes and stretches ... (line 392). During the first six months of gestation, the embryon probably sleeps, as it seems to have no use for voluntary power; it then seems to awake, and to stretch its limbs, and change its posture in some degree, which is termed quickening.

With gills placental ... (line 393). The placenta adheres to any side of the uterus in natural gestation, or of any other cavity in extra-uterine gestation; the extremities of its arteries and veins probably permeate the arteries of the mother, and absorb from thence through their fine coats the oxygen of the mother's blood; hence when the placenta is withdrawn, the side of the uterus, where it adhered, bleeds; but not the extremities of its own vessels.

His dazzled eyes ... (line 398). Though the membrana pupillaris described by modern anatomists guards the tender retina from too much light; the young infant nevertheless seems to feel the presence of it by its frequently moving its eyes, before it can distinguish common objects.

As warmth and moisture ... (line 417).

> In eodem corpore saepe
> Altera pars vivit; rudis est pars altera tellus.
> Quippe ubi temperiem sumpsêre humorque calorque,
> Concipiunt; & ab his oriuntur, cuncta duobus.
> [Often part / alive and part of slime inanimate / are fashioned in one body. Heat combined / with moisture so conceives and life results / from these two things.]
>
> *Ovid: Metamorphoses.*
> *Book I, line 430.*

This story from Ovid of the production of animals from the mud of the

Nile seems to be of Egyptian origin, and is probably a poetical account of the opinions of the magi or priests of that country; showing that the simplest animations were spontaneously produced like chemical combinations, but were distinguished from the latter by their perpetual improvement by the power of reproduction, first by solitary, and then by sexual generation; whereas the products of natural chemistry are only enlarged by accretion, or purified by filtration.

Canto II.
REPRODUCTION OF LIFE

I.

"How short the span of Life! some hours possess'd, *note*
Warm but to cool, and active but to rest!—
The age-worn fibres goaded to contract, *note*
By repetition palsied, cease to act;
When Time's cold hands the languid senses seize,
Chill the dull nerves, the lingering currents freeze;
Organic matter, unreclaim'd by Life,
Reverts to elements by chemic strife.
Thus Heat evolv'd from some fermenting mass
Expands the kindling atoms into gas; 10
Which sink ere long in cold concentric rings,
Condensed, on Gravity's descending wings.

"But Reproduction with ethereal fires *note*
New Life rekindles, ere the first expires;
Calls up renascent Youth, ere tottering age
Quits the dull scene, and gives him to the stage;
Bids on his cheek the rose of beauty blow,
And binds the wreaths of pleasure round his brow;
With finer links the vital chain extends,
And the long line of Being never ends. 20

"Self-moving Engines by unbending springs *note*
May walk on earth, or flap their mimic wings;

In tubes of glass mercurial columns rise,
Or sink, obedient to the incumbent skies;
Or, as they touch the figured scale, repeat
The nice gradations of circumfluent heat.
But Reproduction, when the perfect Elf
Forms from fine glands another like itself,
Gives the true character of life and sense,
And parts the organic from the chemic Ens.—
Where milder skies protect the nascent brood,
And earth's warm bosom yields salubrious food;
Each new Descendant with superior powers
Of sense and motion speeds the transient hours;
Braves every season, tenants every clime,
And Nature rises on the wings of Time.

"As Life discordant elements arrests,
Rejects the noxious, and the pure digests;
Combines with Heat the fluctuating mass,
And gives a while solidity to gas;
Organic forms with chemic changes strive,
Live but to die, and die but to revive!
Immortal matter braves the transient storm,
Mounts from the wreck, unchanging but in form.—

"So, as the sages of the East record
In sacred symbol, or unletter'd word;
Emblem of Life, to change eternal doom'd,
The beauteous form of fair Adonis bloom'd.—
On Syrian hills the graceful Hunter slain
Dyed with his gushing blood the shuddering plain;
And, slow-descending to the Elysian shade,
A while with Proserpine reluctant stray'd;
Soon from the yawning grave the bursting clay
Restor'd the Beauty to delighted day;
Array'd in youth's resuscitated charms,
And young Dione woo'd him to her arms.—

Pleased for a while the assurgent youth above
Relights the golden lamp of life and love;
Ah, soon again to leave the cheerful light,
And sink alternate to the realms of night. 60

II.

"Hence ere Vitality, as time revolves,
Leaves the cold organ, and the mass dissolves;
The Reproductions of the living Ens
From sires to sons, unknown to sex, commence.
New buds and bulbs the living fibre shoots
On lengthening branches, and protruding roots;
Or on the father's side from bursting glands
The adhering young its nascent form expands;
In branching lines the parent-trunk adorns,
And parts ere long like plumage, hairs, or horns. 70

"So the lone Truffle, lodged beneath the earth, *note*
Shoots from paternal roots the tuberous birth;
No stamen-males ascend, and breathe above,
No seed-born offspring lives by female love.
From each young tree, for future buds design'd
Organic drops exsude beneath the rind;
While these with appetencies nice invite, *note*
And those with apt propensities unite;
New embryon fibrils round the trunk combine
With quick embrace, and form the living line: 80
Whose plume and rootlet at their early birth
Seek the dry air, or pierce the humid earth.

"So safe in waves prolific Volvox dwells, *note*
And five descendants crowd his lucid cells;
So the male Polypus parental swims, *note*
And branching infants bristle all his limbs;

So the lone Taenia, as he grows, prolongs　　　*note*
His flatten'd form with young adherent throngs;
Unknown to sex the pregnant oyster swells,　　　*note*
And coral-insects build their radiate shells;　　　*note*
Parturient Sires caress their infant train,　　　91
And heaven-born Storge weaves the social chain;　　　*note*
Successive births her tender cares combine,
And soft affections live along the line.

"On angel-wings the Goddess Form descends,
Round her fond broods her silver arms she bends;
White streams of milk her tumid bosom swell,
And on her lips ambrosial kisses dwell.
Light joys on twinkling feet before her dance
With playful nod, and momentary glance;　　　100
Behind, attendant on the pansied plain,
Young Psyche treads with Cupid in her train.

III.

"In these lone births no tender mothers blend
Their genial powers to nourish or defend;
No nutrient streams from Beauty's orbs improve
These orphan babes of solitary love;
Birth after birth the line unchanging runs,
And fathers live transmitted in their sons;
Each passing year beholds the unvarying kinds,
The same their manners, and the same their minds.　　　110
Till, as erelong successive buds decay,
And insect-shoals successive pass away,
Increasing wants the pregnant parents vex
With the fond wish to form a softer sex;　　　*note*
Whose milky rills with pure ambrosial food
Might charm and cherish their expected brood.
The potent wish in the productive hour

Calls to its aid Imagination's power,   *note*
O'er embryon throngs with mystic charm presides,
And sex from sex the nascent world divides,   120
With soft affections warms the callow trains,
And gives to laughing Love his nymphs and swains;   *note*
Whose mingling virtues interweave at length
The mother's beauty with the father's strength.

"So tulip-bulbs emerging from the seed,
Year after year unknown to sex proceed;
Erewhile the stamens and the styles display
Their petal-curtains, and adorn the day;
The beaux and beauties in each blossom glow
With wedded joy, or amatorial woe.   130
Unmarried Aphides prolific prove
For nine successions uninform'd of love;
New sexes next with softer passions spring,
Breathe the fond vow, and woo with quivering wing.

"So erst in Paradise creation's Lord,
As the first leaves of holy writ record,
From Adam's rib, who press'd the flowery grove,
And dreamt delighted of untasted love,
To cheer and charm his solitary mind,
Form'd a new sex, the Mother of Mankind.   *note*
—Buoy'd on light step the Beauty seem'd to swim,   141
And stretch'd alternate every pliant limb;
Pleased on Euphrates' velvet margin stood,
And view'd her playful image in the flood;
Own'd the fine flame of love, as life began,
And smiled enchantment on adoring Man.
Down her white neck and o'er her bosom roll'd,
Flow'd in sweet negligence her locks of gold;
Round her fine form the dim transparence play'd,
And show'd the beauties, that it seem'd to shade.   150
—Enamour'd Adam gaz'd with fond surprise,

And drank delicious passion from her eyes;
Felt the new thrill of young Desire, and press'd
The graceful Virgin to his glowing breast.—
The conscious Fair betrays her soft alarms,
Sinks with warm blush into his closing arms,
Yields to his fond caress with wanton play,
And sweet, reluctant, amorous, delay.

IV.

"Where no new Sex with glands nutritious feeds,
Nurs'd in her womb, the solitary breeds;   160
No Mother's care their early steps directs,
Warms in her bosom, with her wings protects;
The clime unkind, or noxious food instills
To embryon nerves hereditary ills;
The feeble births acquired diseases chase,   *note*
Till Death extinguish the degenerate race.

"So grafted trees with shadowy summits rise,   *note*
Spread their fair blossoms, and perfume the skies;
Till canker taints the vegetable blood,
Mines round the bark, and feeds upon the wood.   170
So, years successive, from perennial roots
The wire or bulb with lessen'd vigour shoots;
Till curled leaves, or barren flowers, betray
A waning lineage, verging to decay;
Or till, amended by connubial powers,
Rise seedling progenies from sexual flowers.

"E'en where unmix'd the breed, in sexual tribes
Parental taints the nascent babe imbibes;
Eternal war the Gout and Mania wage
With fierce uncheck'd hereditary rage;   180

Sad Beauty's form foul Scrofula surrounds
With bones distorted, and putrescent wounds;
And, fell Consumption! thy unerring dart   *note*
Wets its broad wing in Youth's reluctant heart.

"With pausing step, at night's refulgent noon,
Beneath the sparkling stars, and lucid moon,
Plung'd in the shade of some religious tower,
The slow bell counting the departed hour,
O'er gaping tombs where shed umbrageous Yews
On mouldering bones their cold unwholesome dews;   190
While low aerial voices whisper round,
And moondrawn spectres dance upon the ground;
Poetic Melancholy loves to tread,
And bend in silence o'er the countless Dead;
Marks with loud sobs infantine Sorrows rave,
And wring their pale hands o'er their Mother's grave;
Hears on the new-turn'd sod with gestures wild
The kneeling Beauty call her buried child;
Upbraid with timorous accents Heaven's decrees,
And with sad sighs augment the passing breeze.   200
'Stern Time,' She cries, 'receives from Nature's womb
Her beauteous births, and bears them to the tomb;
Calls all her sons from earth's remotest bourn,
And from the closing portals none return!'

V.

Urania paused,—upturn'd her streaming eyes,
And her white bosom heaved with silent sighs;
With her the Muse laments the sum of things,
And hides her sorrows with her meeting wings;
Long o'er the wrecks of lovely Life they weep,
Then pleased reflect, "to die is but to sleep;"   210
From Nature's coffins to her cradles turn,

Smile with young joy, with new affection burn.

And now the Muse, with mortal woes impress'd,
Thus the fair Hierophant again address'd.
—"Ah me! celestial Guide, thy words impart
Ills undeserved, that rend the nascent heart!
O, Goddess, say, if brighter scenes improve
Air-breathing tribes, and births of sexual love?"—
The smiling Fair obeys the inquiring Muse,
And in sweet tones her grateful task pursues. 220

"Now on broad pinions from the realms above
Descending Cupid seeks the Cyprian grove;
To his wide arms enamour'd Psyche springs, *note*
And clasps her lover with aurelian wings.
A purple sash across His shoulder bends,
And fringed with gold the quiver'd shafts suspends;
The bending bow obeys the silken string,
And, as he steps, the silver arrows ring.
Thin folds of gauze with dim transparence flow
O'er Her fair forehead, and her neck of snow; 230
The winding woof her graceful limbs surrounds,
Swells in the breeze, and sweeps the velvet grounds;
As hand in hand along the flowery meads
His blushing bride the quiver'd hero leads;
Charm'd round their heads pursuing Zephyrs throng,
And scatter roses, as they move along;
Bright beams of Spring in soft effusion play,
And halcyon Hours invite them on their way.

"Delighted Hymen hears their whisper'd vows,
And binds his chaplets round their polish'd brows, 240
Guides to his altar, ties the flowery bands,
And as they kneel, unites their willing hands.
'Behold, he cries, Earth! Ocean! Air above,
'And hail the Deities of Sexual Love!

'All forms of Life shall this fond Pair delight,
'And sex to sex the willing world unite;
'Shed their sweet smiles in Earth's unsocial bowers,
'Fan with soft gales, and gild with brighter hours;
'Fill Pleasure's chalice unalloy'd with pain,
'And give Society his golden chain.'   250

"Now young Desires, on purple pinions borne,
Mount the warm gales of Manhood's rising morn;
With softer fires through virgin bosoms dart,
Flush the pale cheek, and goad the tender heart.
Ere the weak powers of transient Life decay,
And Heaven's ethereal image melts away;
Love with nice touch renews the organic frame,
Forms a young Ens, another and the same;
Gives from his rosy lips the vital breath,
And parries with his hand the shafts of death;   260
While Beauty broods with angel wings unfurl'd   *note*
O'er nascent life, and saves the sinking world.

"Hence on green leaves the sexual Pleasures dwell,
And Loves and Beauties crowd the blossom's bell;
The wakeful Anther in his silken bed
O'er the pleased Stigma bows his waxen head;
With meeting lips and mingling smiles they sup
Ambrosial dewdrops from the nectar'd cup;   *note*
Or buoy'd in air the plumy Lover springs,
And seeks his panting bride on Hymen-wings.   270

"The Stamen males, with appetencies just,   *note*
Produce a formative prolific dust;
With apt propensities, the Styles recluse
Secrete a formative prolific juice;
These in the pericarp erewhile arrive,
Rush to each other, and embrace alive.
—Form'd by new powers progressive parts succeed,

Join in one whole, and swell into a seed.

"So in fond swarms the living Anthers shine
Of bright Vallisner on the wavy Rhine;   *note*
Break from their stems, and on the liquid glass   281
Surround the admiring stigmas as they pass;
The love-sick Beauties lift their essenced brows,
Sigh to the Cyprian queen their secret vows,
Like watchful Hero feel their soft alarms,
And clasp their floating lovers in their arms.

"Hence the male Ants their gauzy wings unfold,
And young Lampyris waves his plumes of gold;   *note*
The Glow-Worm sparkles with impassion'd light
On each green bank, and charms the eye of night;   290
While new desires the painted Snail perplex,
And twofold love unites the double sex.

"Hence, when the Morus in Italia's lands
To spring's warm beam its timid leaf expands;
The Silk-Worm broods in countless tribes above
Crop the green treasure, uninform'd of love;
Erewhile the changeful worm with circling head
Weaves the nice curtains of his silken bed;
Web within web involves his larva form,
Alike secured from sunshine and from storm;   300
For twelve long days He dreams of blossom'd groves,
Untasted honey, and ideal loves;   *note*
Wakes from his trance, alarm'd with young Desire,
Finds his new sex, and feels ecstatic fire;
From flower to flower with honey'd lip he springs,
And seeks his velvet loves on silver wings.

VI.

"The Demon, Jealousy, with Gorgon frown
Blasts the sweet flowers of Pleasure not his own,
Rolls his wild eyes, and through the shuddering grove
Pursues the steps of unsuspecting Love;   310
Or drives o'er rattling plains his iron car,
Flings his red torch, and lights the flames of war.

Here Cocks heroic burn with rival rage,
And Quails with Quails in doubtful fight engage;
Of armed heels and bristling plumage proud,
They sound the insulting clarion shrill and loud,
With rustling pinions meet, and swelling chests,
And seize with closing beaks their bleeding crests;
Rise on quick wing above the struggling foe,
And aim in air the death-devoting blow.   320
There the hoarse stag his croaking rival scorns,   *note*
And butts and parries with his branching horns;
Contending Boars with tusk enamell'd strike,
And guard with shoulder-shield the blow oblique;
While female bands attend in mute surprise,
And view the victor with admiring eyes.—

"So Knight on Knight, recorded in romance,
Urged the proud steed, and couch'd the extended lance;
He, whose dread prowess with resistless force,
O'erthrew the opposing warrior and his horse,   330
Bless'd, as the golden guerdon of his toils,
Bow'd to the Beauty, and receiv'd her smiles.

"So when fair Helen with ill-fated charms,
By Paris wooed, provoked the world to arms,

Left her vindictive Lord to sigh in vain
For broken vows, lost love, and cold disdain;
Fired at his wrongs, associate to destroy
The realms unjust of proud adulterous Troy,
Unnumber'd Heroes braved the dubious fight,
And sunk lamented to the shades of night.   340

"Now vows connubial chain the plighted pair,
And join paternal with maternal care;
The married birds with nice selection cull
Soft thistle-down, gray moss, and scattered wool,
Line the secluded nest with feathery rings,
Meet with fond bills, and woo with fluttering wings.
Week after week, regardless of her food,
The incumbent Linnet warms her future brood;   *note*
Each spotted egg with ivory lips she turns,
Day after day with fond expectance burns,   350
Hears the young prisoner chirping in his cell,   *note*
And breaks in hemispheres the obdurate shell.
Loud trills sweet Philomel his tender strain,
Charms his fond bride, and wakes his infant train;
Perch'd on the circling moss, the listening throng
Wave their young wings, and whisper to the song.   *note*

"The Lion-King forgets his savage pride,
And courts with playful paws his tawny bride;
The listening Tiger hears with kindling flame
The love-lorn night-call of his brinded dame.   360
Despotic Love dissolves the bestial war,
Bends their proud necks, and joins them to his car;
Shakes o'er the obedient pairs his silken thong,
And goads the humble, or restrains the strong.—
Slow roll the silver wheels,—in beauty's pride
Celestial Psyche blushing by his side.—
The lordly Bull behind and warrior Horse
With voice of thunder shake the echoing course,

Chain'd to the car with herds domestic move,
And swell the triumph of despotic Love.   370

"Pleased as they pass along the breezy shore
In twinkling shoals the scaly realms adore,
Move on quick fin with undulating train,   *note*
Or lift their slimy foreheads from the main.
High o'er their heads on pinions broad display'd   *note*
The feather'd nations shed a floating shade;
Pair after pair enamour'd shoot along,
And trill in air the gay impassion'd song.
With busy hum in playful swarms around
Emerging insects leave the peopled ground,   380
Rise in dark clouds, and borne in airy rings
Sport round the car, and wave their golden wings.
Admiring Fawns pursue on dancing hoof,
And bashful Dryads peep from shades aloof;
Emerging Nereids rise from coral cells,
Enamour'd Tritons sound their twisted shells;
From sparkling founts enchanted Naiads move,
And swell the triumph of despotic Love.

"Delighted Flora, gazing from afar,
Greets with mute homage the triumphal car;   390
On silvery slippers steps with bosom bare,
Bends her white knee, and bows her auburn hair;
Calls to her purple heaths, and blushing bowers,
Bursts her green gems, and opens all her flowers;
O'er the bright Pair a shower of roses sheds,
And crowns with wreathes of hyacinth their heads.—
—Slow roll the silver wheels with snowdrops deck'd,
And primrose bands the cedar spokes connect;
Round the fine pole the twisting woodbine clings,
And knots of jasmine clasp the bending springs;   400
Bright daisy links the velvet harness chain,
And rings of violets join each silken rein;

Festoon'd behind, the snow-white lilies bend,
And tulip-tassels on each side depend.
—Slow rolls the car,—the enamour'd Flowers exhale
Their treasured sweets, and whisper to the gale;
Their ravelled buds, and wrinkled cups unfold,
Nod their green stems, and wave their bells of gold;
Breathe their soft sighs from each enchanted grove,
And hail The Deities of Sexual Love. 410

"Onward with march sublime in saffron robe
Young Hymen steps, and traverses the globe;
O'er burning sands, and snow-clad mountains, treads,
Blue fields of air, and ocean's briny beds;
Flings from his radiant torch celestial light
O'er Day's wide concave, and illumes the Night.
With dulcet eloquence his tuneful tongue
Convokes and captivates the Fair and Young;
His golden lamp with ray ethereal dyes
The blushing cheek, and lights the laughing eyes; 420
With secret flames the virgin's bosom warms,
And lights the impatient bridegroom to her arms;
With lovely life all Nature's frame inspires,
And, as they sink, rekindles all her fires."

VII.

Now paused the beauteous Teacher, and awhile
Gazed on her train with sympathetic smile.
'Beware of Love! she cried, ye Nymphs, and hear
'His twanging bowstring with alarmed ear;
'Fly the first whisper of the distant dart,
'Or shield with adamant the fluttering heart; 430
'To secret shades, ye Virgin trains, retire,
'And in your bosoms guard the vestal fire.'
—The obedient Beauties hear her words, advised,

And bow with laugh repress'd, and smile chastised.   *notes*

Now at her nod the Nymphs attendant bring
Translucent water from the bubbling spring;
In crystal cups the waves salubrious shine,
Unstain'd untainted with immodest wine.
Next, where emerging from its ancient roots
Its widening boughs the Tree of Knowledge shoots;   440
Pluck'd with nice choice before the Muse they placed
The now no longer interdicted taste.
Awhile they sit, from higher cares released,
And pleased partake the intellectual feast.
Of good and ill they spoke, effect and cause,
Celestial agencies, and Nature's laws.

So when angelic Forms to Syria sent
Sat in the cedar shade by Abraham's tent;
A spacious bowl the admiring Patriarch fills
With dulcet water from the scanty rills;   450
Sweet fruits and kernels gathers from his hoard,
With milk and butter piles the plenteous board;
While on the heated hearth his Consort bakes
Fine flour well kneaded in unleaven'd cakes.
The Guests ethereal quaff the lucid flood,
Smile on their hosts, and taste terrestrial food;
And while from seraph-lips sweet converse springs,
Lave their fair feet, and close their silver wings.

End of Canto II.

NOTES TO CANTO II.

How short the span of Life ... (line 1). The thinking few in all ages have complained of the brevity of life, lamenting that mankind are not allowed time sufficient to cultivate science, or to improve their intellect. Hippocrates introduces his celebrated aphorisms with this idea; "Life is short, science long, opportunities of knowledge rare, experiments fallacious, and reasoning difficult."—A melancholy reflection to philosophers!

The age-worn fibres ... (line 3). Why the same kinds of food, which enlarge and invigorate the body from infancy to the meridian of life, and then nourish it for some years unimpaired, should at length gradually cease to do so, and the debility of age and death supervene, would be liable to surprise us if we were not in the daily habit of observing it; and is a circumstance which has not yet been well understood.

Before mankind introduced civil society, old age did not exist in the world, nor other lingering diseases; as all living creatures, as soon as they became too feeble to defend themselves, were slain and eaten by others, except the young broods, who were defended by their mother; and hence the animal world existed uniformly in its greatest strength and perfection; see Additional Note VII.

But Reproduction ... (line 13). See Additional Note VIII.

Unbending springs ... (line 21). See Additional Note I, 4.

Combines with Heat ... (line 39). It was shown in note on line 248 of the first Canto, that much of the aerial and liquid parts of the terraqueous globe was converted by the powers of life into solid matter; and that this was effected by the combination of the fluid, heat, with other elementary bodies by the appetencies and propensities of the parts of living matter to unite with each other. But when these appetencies and propensities of the parts of organic matter to unite with each other cease, the chemical affinities of attraction and the aptitude to be attracted, and of repulsion and the aptitude to be repelled, succeed, and reduce much of the solid matters back to the condition of elements; which seems to be effected by the matter of heat being again set at liberty, which was combined with other matters by the powers of life; and thus by its diffusion the solid bodies return into liquid ones or into gasses, as occurs in the processes of fermentation, putrefaction, sublimation, and calcination. Whence solidity appears to be produced in consequence of the diminution of heat, as the condensation of steam into water, and the consolidation of water into ice, or by the combination of heat with bodies, as with the materials of gunpowder before its explosion.

Immortal matter ... (line 43). The perpetual mutability of the forms of matter seems to have struck the philosophers of great antiquity; the system of transmigration taught by Pythagoras, in which the souls of men were supposed after death to animate the bodies of a variety of animals, appears to have arisen from this source. He had observed the perpetual changes of organic matter from one creature to another, and concluded, that the vivifying spirit must attend it.

Emblem of Life ... (line 47). The Egyptian figure of Venus rising from the sea seems to have represented the Beauty of organic Nature; which the philosophers of that country, the magi, appear to have discovered to have been elevated by earthquakes from the primeval ocean. But the hieroglyphic figure of Adonis seems to have signified the spirit of animation or life, which was perpetually wooed or courted by organic matter, and which perished and revived alternately. Afterwards the fable of Adonis seems to have given origin to the first religion promising a resurrection from the dead; whence his funeral and return to life were celebrated for many ages in Egypt and Syria, the ceremonies of which Ezekiel complains as idolatrous, accusing the women of Israel of lamenting over Thammus; which St. Cyril interprets to be Adonis, in his Commentaries on Isaiah; Danet's Dictionary.

So the lone Truffle ... (line 71). Lycoperdon tuber. This plant never rises above the earth, is propagated without seed by its roots only, and seems to require no light. Perhaps many other fungi are generated without seed by their roots only, and without light, and approach on the last account to animal nature.

While these with appetencies ... (line 77). See Additional Note VIII.

Prolific Volvox ... (line 83). The Volvox globator dwells in the lakes of Europe, is transparent, and bears within it children and grandchildren to the fifth generation; Systema Naturae.

The male polypus ... (line 85). The Hydra viridis and fusca of Linneus dwell in our ditches and rivers under aquatic plants; these animals have been shown by ingenious observers to revive after having been dried, to be restored when mutilated, to be multiplied by dividing them, and propagated from portions of them, parts of different ones to unite, to be turned inside outwards and yet live, and to be propagated by seeds, to produce bulbs, and vegetate by branches. Systema Naturae.

The lone Taenia ... (line 87). The tape-worm dwells in the intestines of animals, and grows old at one extremity, producing an infinite series of young ones at the other; the separate joints have been called Gourd-worms, each of which possesses a mouth of its own, and organs of digestion. Systema Naturae.

The pregnant oyster ... (line 89). Ostrea edulis dwells in the European oceans, frequent at the tables of the luxurious, a living repast! New-born oysters swim swiftly by an undulating movement of fins thrust out a little way from their shells. Systema Naturae. But they do not afterwards change their place during their whole lives, and are capable of no other movement but that of opening the shell a little way: whence Professor Beckman observes, that their offspring is probably produced without maternal organs; and that those, who speak of male and female oysters, must be mistaken: Philosophical Magazine, March 1800. It is also observed by H.I. le Beck, that on nice inspection of the Pearl oysters in the gulf of Manar, he could observe no distinction of sexes. Nicholson's Journal, April 1800.

And coral insects ... (line 90). The coral habitation of the Madrepora of Linneus consists of one or more star-like cells; a congeries of which form rocks beneath the sea; the animal which constructs it is termed Medusa; and as it adheres to its calcareous cavity, and thence cannot travel to its neighbours, is probably without sex. I observed great masses of the limestone in Shropshire, which is brought to Newport, to consist of the cells of these animals.

And heaven-born Storge ... (line 92). See Additional Note IX.

A softer sex ... (line 114). The first buds of trees raised from seed die annually, and are succeeded by new buds by solitary reproduction; which are larger or more perfect for several successive years, and then they produce sexual flowers, which are succeeded by seminal reproduction. The same occurs in bulbous rooted plants raised from seed; they die annually, and produce others rather more perfect than the parent for several years, and then produce sexual flowers. The Aphis is in a similar manner hatched from an egg in the vernal months, and produces a viviparous offspring without sexual intercourse for nine or ten successive generations; and then the progeny is both male and female, which cohabit, and from these new females are produced eggs, which endure the winter; the same process probably occurs in many other insects.

Imagination's power ... (line 118). The manner in which the similarity of the progeny to the parent, and the sex of it, are produced by the power of imagination, is treated of in Zoonomia, Sect. XXXIX, 6, 3. It is not to be understood, that the first living fibres, which are to form an animal, are produced by imagination, with any similarity of form to the future animal; but with appetencies or propensities, which shall produce by accretion of parts the similarity of form and feature, or of sex, corresponding with the imagination of the father.

His nymphs and swains ... (line 122). The arguments which have been ad-duced to show, that mankind and quadrupeds were formerly in an hermaphrodite state, are first deduced from the present existence of breasts and nipples in all the males; which latter swell on titillation like those of the females, and which are said to contain a milky fluid at their birth; and it is affirmed, that some men have given milk to their children in desert countries, where the mother has perished; as the male pigeon is said to give a kind of milk from his stomach along with the regurgitated food, to the young doves, as mentioned in Additional Note IX, on Storge.

Secondly, from the apparent progress of many animals to greater perfection, as in some insects, as the flies with two wings, termed Diptera; which have rudiments of two other wings, called halteres, or poisers; and in many flowers which have rudiments of new stamina, or filaments without anthers on them. See Botanic Garden, Vol. II, Curcuma, Note, and the Note on line 204 of Canto I of this work. It has been supposed by some, that mankind were formerly quadrupeds as well as hermaphrodites; and that some parts of the body are not yet so convenient to an erect attitude as to a horizontal one; as the fundus of the bladder in an erect posture is not exactly over the insertion of the urethra; whence it is seldom completely evacuated, and thus renders mankind more subject to the stone, than if he had preserved his horizontality: these philosophers, with Buffon and Helvetius, seem to imagine, that mankind arose from one family of monkeys on the banks of the Mediterranean; who accidentally had learned to use the adductor pollicis, or that strong muscle which constitutes the ball of the thumb, and draws the point of it to meet the points of the fingers; which common monkeys do not; and that this muscle gradually increased in size, strength, and activity, in successive generations; and by this improved use of the sense of touch, that monkeys acquired clear ideas, and gradually became men.

Perhaps all the productions of nature are in their progress to greater perfection! an idea countenanced by modern discoveries and deductions concerning the progressive formation of the solid parts of the terraqueous globe, and consonant to the dignity of the Creator of all things.

The mother of mankind ... (line 140). See Additional Note X.

Acquired diseases ... (line 165). See Additional Note XI.

So grafted trees ... (line 167). Mr. Knight first observed that those apple and pear trees, which had been propagated for above a century by ingraftment were now so unhealthy, as not to be worth cultivation. I have suspected the diseases of potatoes attended with the curled leaf, and of strawberry plants attended with barren flowers, to be owing to their having been too long raised from roots, or by solitary reproduction, and not from seeds, or sexual reproduction, and to have thence acquired those hereditary diseases.

And, fell Consumption ... (line 183).

... Haeret lateri lethalis arundo.
[... The fatal dart sticks in her side.]
Virgil.

Enamoured Psyché ... (line 223). A butterfly was the ancient emblem of the soul after death as rising from the tomb of its former state, and becoming a winged inhabitant of air from an insect creeping upon earth. At length the wings only were given to a beautiful nymph under the name of Psyche, which is the greek word for the soul, and also became afterwards to signify a butterfly probably from the popularity of this allegory. Many allegorical designs of Cupid or Love warming a butterfly or the Soul with his torch may be seen in Spence's Polymetis, and a beautiful one of their marriage in Bryant's Mythology; from which this description is in part taken.

While Beauty broods ... (line 261).

Alma Venus! [...] per te quoniam genus omne animantum
[...] Concipitur, visitque exortum lumina [...] coeli.
Lucretius.

[*Editors note:* The deleted parts of the text are not marked in the original. The complete text reads as this: "alma Venus, caeli subter labentia signa / quae mare navigerum, quae terras frugiferentis / concelebras, per te quoniam genus omne animantum / concipitur visitque exortum lumina solis: / te, dea, te fugiunt venti, te nubila caeli." (De Rerum Natura, Book I.)

In W.E. Leonard's translation: "Dear Venus that beneath the gliding stars / Makest to teem the many-voyaged main / And fruitful lands— for all of living things / Through thee alone are evermore conceived / Through thee are risen to visit the great sun / Before thee, Goddess, and thy coming on / Flee stormy wind and massy cloud away."]

From the nectar'd cup ... (line 268). The anthers and stigmas of flowers are probably nourished by the honey, which is secreted by the honey-gland called by Linneus the nectary; and possess greater sensibility or animation than other parts of the plant. The corol of the flower appears to be a respiratory organ belonging to these anthers and stigmas for the purpose of further oxygenating the vegetable blood for the production of the anther dust and of this honey, which is also exposed to the air in its receptacle or honey-cup; which, I suppose, to be necessary for its further oxygenation, as in many flowers so complicate an apparatus is formed for its protection from insects, as in aconitum, delphinium, larkspur, lonicera, woodbine; and because the corol and nectary fall along with the anthers and stigmas, when the pericarp is impregnated.

Dr. B. S. Barton in the American Transactions has lately shown, that the honey collected from some plants is intoxicating and poisonous to men, as from rhododendron, azalea, and datura; and from some other plants that it is hurtful to the bees which collect it; and that from some flowers it is so injurious or disagreeable, that they do not collect it, as from the fritillaria or crown imperial of this country.

With appetencies just ... (line 271). As in the productions by chemical affinity one set of particles must possess the power of attraction, and the other the aptitude to be attracted, as when iron approaches a magnet; so when animal particles unite, whether in digestion or reproduction, some of them must possess an appetite to unite, and others a propensity to be united. The former of these are secreted by the anthers from the vegetable blood, and the latter by the styles or pericarp; see the Additional Note VIII, on Reproduction.

Of bright Vallisner ... (line 280). Vallisneria, of the class of dioecia. The flowers of the male plant are produced under water, and as soon as their farina or dust is mature, they detach themselves from the plant, rise to the surface and continue to flourish, and are wafted by the air or borne by the current to the female flowers. In this they resemble those tribes of insects, where the males at certain seasons acquire wings, but not the females, as ants, coccus, lampyris, phalaena, brumata, lichanella; Botanic Garden, Vol. II, Note on Vallisneria.

And young Lampyris ... (line 288). The fire-fly is at some seasons so luminous, that M. Merian says, that by putting two of them under a glass, she was able to draw her figures of them by night. Whether the light of this and of other insects be caused by their amatorial passion, and thus assists them to find each other; or is caused by respiration, which is so analogous to combustion; or to a tendency to putridity, as in dead fish and rotten wood, is still to be investigated; see Botanic Garden, Vol. I, Additional Note IX.

Untasted honey ... (line 302). The numerous moths and butterflies seem to pass from a reptile leaf-eating state, and to acquire wings to flit in air, with a proboscis to gain honey for their food along with their organs of reproduction, solely for the purpose of propagating their species by sexual intercourse, as they die when that is completed. By the use of their wings they have access to each other on different branches or on different vegetables, and by living upon honey probably acquire a higher degree of animation, and thus seem to resemble the anthers of flowers, which probably are supported by honey only, and thence acquire greater sensibility; see Note on Vallisneria, line 280 of this Canto.

A naturalist, who had studied this subject, thought it not impossible that the first insects were the anthers and stigmas of flowers, which had by some means loosened themselves from their parent plant, like the male flowers of vallisneria, and that other insects in process of time had been formed from these, some acquiring wings, others fins, and others claws, from their ceaseless efforts to procure food or to secure themselves from injury. He contends, that none of these changes are more incomprehensible than the transformation of caterpillars into butterflies; see Botanic Garden, Vol. I, Additional Note XXXIX.

There the hoarse stag ... (line 321). A great want of one part of the animal world has consisted in the desire of the exclusive possession of the females; and these have acquired weapons to combat each other for this purpose, as the very thick shield-like horny skin on the shoulder of the boar is a defence only against animals of his own species, who strike obliquely upwards, nor are his tushes for other purposes, except to defend himself, as he is not naturally a carnivorous animal. So the horns of the stag are sharp to offend his adversary, but are branched for the purpose of parrying or receiving the thrusts of horns similar to his own, and have therefore been formed for the purpose of combating other stags for the exclusive possession of the females, who are observed, like the ladies in the times of chivalry, to attend the car of the victor.

The birds, which do not carry food to their young, and do not therefore marry, are armed with spurs for the purpose of fighting for the exclusive possession of the females, as cocks and quails. It is certain that these weapons are not provided for their defence against other adversaries, because the females of these species are without this armour; Zoonomia, Sect. XXXIX, 4, 8.

The incumbent Linnet ... (line 348). The affection of the unexperienced and untaught bird to its egg, which induces it to sit days and weeks upon it to warm the enclosed embryon, is a matter of great difficulty to explain; See Additional Note IX, on Storge. Concerning the fabrication of their nests, see Zoonomia, Sect. XVI, 13, on instinct.

Hears the young prisoner ... (line 351). The air-vessel at the broad end of an incubated egg gradually extends its edges along the sides of the shell, as the chick enlarges, but is at the same time applied closer to the internal surface of the shell; when the time of hatching approaches the chick is liable to break this air-bag with its beak, and thence begin to breathe and to chirp; at this time the edges of the enlarged air-bag extend so as to cover internally one hemisphere of the egg; and as one half of the external shell is thus moist, and the other half dry, as soon as the mother hearing the chick chirp, or the chick itself wanting respirable air, strikes the egg, about its equatorial line, it breaks into two hemispheres, and liberates the prisoner.

And whisper to the song ... (line 356). A curious circumstance is mentioned by Kircherus de Musurgia, in his Chapter de Lusciniis. "That the young nightingales, that are hatched under other birds, never sing till they are instructed by the company of other nightingales." And Johnston affirms, that the nightingales that visit Scotland, have not the same harmony as those of Italy, (Pennant's Zoology, octavo, p. 255), which would lead us to suspect, that the singing of birds, like human music, is an artificial language rather than a natural expression of passion.

With undulating train ... (line 373). The side fins of fish seem to be chiefly used to poise them; as they turn upon their backs immediately when killed, the air-bladder assists them perhaps to rise or descend by its possessing the power to condense the air in it by muscular contraction; and it is possible, that at great depths in the ocean the air in this receptacle may by the great pressure of the incumbent water become condensed into so small a space, as to cease to be useful to the animal, which was possibly the cause of the death of Mr. Day in his diving ship. See note on Ulva, Botanic Garden, Vol. II.

The progressive motion of fish beneath the water is produced principally by the undulation of their tails. One oblique plain of a part of the tail on the right side of the fish strikes the water at the same time that another oblique plain strikes it on the left side, hence in respect to moving to the right or left these percussions of the water counteract each other, but they coincide in respect to the progression of the fish; this power seems to be better applied to push forwards a body in water, than the oars of boats, as the particles of water recede from the stroke of the oar, whence the comparative power acquired is but as the difference of velocity between the striking oar and the receding water. So a ship moves swifter with an oblique wind, than with a wind of the same velocity exactly behind it; and the common windmill sail placed obliquely to the wind is more powerful than one which directly recedes from it. Might not some machinery resembling the tails of fish be placed behind a boat, so as to be moved with greater effect than common oars, by the force of wind or steam, or perhaps by hand?

On pinions broad display'd ... (line 375). The progressive motion of birds in the air is principally performed by the movement of their wings, and not by that of their tails as in fish. The bird is supported in an element so much lighter than itself by the resistance of the air as it moves horizontally against the oblique plain made by its breast, expanded tail and wings, when they are at rest; the change of this obliquity also assists it to rise, and even directs its descent, though this is owing principally to its specific gravity, but it is in all situations kept upright or balanced by its wings.

As the support of the bird in the air, as well as its progression, is performed by the motion of the wings; these require strong muscles as are seen on the breasts of partridges. Whence all attempts of men to fly by wings applied to the weak muscles of their arms have been ineffectual; but it is not certain whether light machinery so contrived as to be moved by their feet, might not enable them to fly a little way, though not so as to answer any useful purpose.

With laugh repress'd ... (line 434). The cause of the violent actions of laughter, and of the difficulty of restraining them, is a curious subject of inquiry. When pain afflicts us, which we cannot avoid, we learn to relieve it by great voluntary exertions, as in grinning, holding the breath, or screaming; now the pleasurable sensation, which excites laughter, arises for a time so high as to change its name, and become a painful one; and we excite the convulsive motions of the respiratory muscles to relieve this pain. We are however unwilling to lose the pleasure, and presently put a stop to this exertion; and immediately the pleasure recurs, and again as instantly rises into pain. Which is further explained in Zoonomia, Sect. XXXIV, 1, 4. When this pleasurable sensation rises into a painful one, and the customs of society will not permit us to laugh aloud, some other violent voluntary exertion is used instead of it to alleviate the pain.

With smile chastised ... (line 434). The origin of the smile has generally been ascribed to inexplicable instinct, but may be deduced from our early associations of actions and ideas. In the act of sucking, the lips of the infant are closed round the nipple of its mother, till it has filled its stomach, and the pleasure of digesting this grateful food succeeds; then the sphincter of the mouth, fatigued by the continued action of sucking, is relaxed; and the antagonist muscles of the face gently acting, produce the smile of pleasure, which is thus during our lives associated with gentle pleasure, which is further explained in Zoonomia, Sect. XVI, 8, 4.

Canto III.
PROGRESS OF THE MIND

I.

Now rose, adorn'd with Beauty's brightest hues,
The graceful Hierophant, and winged Muse;
Onward they step around the stately piles,
O'er porcelain floors, through laqueated ailes,
Eye Nature's lofty and her lowly seats,
Her gorgeous palaces, and green retreats,
Pervade her labyrinths with unerring tread,
And leave for future guests a guiding thread.

First with fond gaze blue fields of air they sweep,
Or pierce the briny chambers of the deep;   10
Earth's burning line, and icy poles explore,
Her fertile surface, and her caves of ore;
Or mark how Oxygen with Azote-Gas   *note*
Plays round the globe in one aerial mass,
Or fused with Hydrogen in ceaseless flow
Forms the wide waves, which foam and roll below.

Next with illumined hands through prisms bright
Pleased they untwist the sevenfold threads of light;
Or, bent in pencils by the lens, convey
To one bright point the silver hairs of Day.   20
Then mark how two electric streams conspire   *note*

To form the resinous and vitreous fire;
Beneath the waves the fierce Gymnotus arm,
And give Torpedo his benumbing charm;
Or, through Galvanic chain-work as they pass,
Convert the kindling water into gas.

How at the poles opposing Ethers dwell,
Attract the quivering needle, or repel.
How Gravitation by immortal laws
Surrounding matter to a centre draws; 30
How Heat, pervading oceans, airs, and lands,
With force uncheck'd the mighty mass expands;
And last how born in elemental strife
Beam'd the first spark, and lighten'd into Life.

Now in sweet tones the inquiring Muse express'd
Her ardent wish; and thus the Fair address'd.
"Priestess of Nature! whose exploring sight
Pierces the realms of Chaos and of Night;
Of space unmeasured marks the first and last,
Of endless time the present, future, past; 40
Immortal Guide! O, now with accents kind
Give to my ear the progress of the Mind.
How loves, and tastes, and sympathies commence
From evanescent notices of sense?
How from the yielding touch and rolling eyes
The piles immense of human science rise?—
With mind gigantic steps the puny Elf,
And weighs and measures all things but himself!"

The indulgent Beauty hears the grateful Muse,
Smiles on her pupil, and her task renews. 50
Attentive Nymphs in sparkling squadrons throng,
And choral Virgins listen to the song;
Pleased Fawns and Naiads crowd in silent rings,

And hovering Cupids stretch their purple wings.

II.

"First the new actions of the excited sense,
Urged by appulses from without, commence;
With these exertions pain or pleasure springs,
And forms perceptions of external things.
Thus, when illumined by the solar beams,
Yon waving woods, green lawns, and sparkling streams,   60
In one bright point by rays converging lie
Plann'd on the moving tablet of the eye;
The mind obeys the silver goads of light,
And Irritation moves the nerves of sight.   *note*

"These acts repeated rise from joys or pains,
And swell Imagination's flowing trains;
So in dread dreams amid the silent night
Grim spectre-forms the shuddering sense affright;
Or Beauty's idol-image, as it moves,
Charms the closed eye with graces, smiles, and loves;   70
Each passing form the pausing heart delights,
And young Sensation every nerve excites.   *note*

"Oft from sensation quick Volition springs,   *note*
When pleasure thrills us, or when anguish stings;
Hence Recollection calls with voice sublime
Immersed ideas from the wrecks of Time,
With potent charm in lucid trains displays
Eventful stories of forgotten days.
Hence Reason's efforts good with ill contrast,
Compare the present, future, and the past;   80
Each passing moment, unobserved restrain   *note*
The wild discordancies of Fancy's train;

But leave uncheck'd the Night's ideal streams,
Or, sacred Muses! your meridian dreams.

"And last Suggestion's mystic power describes
Ideal hosts arranged in trains or tribes.
So when the Nymph with volant finger rings
Her dulcet harp, and shakes the sounding strings;
As with soft voice she trills the enamour'd song,
Successive notes, unwill'd, the strain prolong;   90
The transient trains Association steers,   *note*
And sweet vibrations charm the astonish'd ears.

"On rapid feet o'er hills, and plains, and rocks,
Speed the scared leveret and rapacious fox;
On rapid pinions cleave the fields above
The hawk descending, and escaping dove;
With nicer nostril track the tainted ground
The hungry vulture, and the prowling hound;
Converge reflected light with nicer eye
The midnight owl, and microscopic fly;   100
With finer ear pursue their nightly course
The listening lion, and the alarmed horse.

"The branching forehead with diverging horns   *note*
Crests the bold bull, the jealous stag adorns;
Fierce rival boars with side-long fury wield
The pointed tusk, and guard with shoulder-shield;
Bounds the dread tiger o'er the affrighted heath
Arm'd with sharp talons, and resistless teeth;
The pouncing eagle bears in clinched claws
The struggling lamb, and rends with ivory jaws;   110
The tropic eel, electric in his ire,   *note*
Alarms the waves with unextinguish'd fire;
The fly of night illumes his airy way,   *note*
And seeks with lucid lamp his sleeping prey;

Fierce on his foe the poisoning serpent springs,
And insect armies dart their venom'd stings.

"Proud Man alone in wailing weakness born,
No horns protect him, and no plumes adorn;
No finer powers of nostril, ear, or eye,
Teach the young Reasoner to pursue or fly.— 120
Nerved with fine touch above the bestial throngs,
The hand, first gift of Heaven! to man belongs; *note*
Untipt with claws the circling fingers close,
With rival points the bending thumbs oppose, *note*
Trace the nice lines of Form with sense refined,
And clear ideas charm the thinking mind.
Whence the fine organs of the touch impart
Ideal figure, source of every art;
Time, motion, number, sunshine or the storm,
But mark varieties in Nature's form. 130

"Slow could the tangent organ wander o'er
The rock-built mountain, and the winding shore;
No apt ideas could the pigmy mite,
Or embryon emmet to the touch excite;
But as each mass the solar ray reflects,
The eye's clear glass the transient beams collects;
Bends to their focal point the rays that swerve,
And paints the living image on the nerve.
So in some village-barn, or festive hall
The spheric lens illumes the whiten'd wall; 140
O'er the bright field successive figures fleet,
And motley shadows dance along the sheet.—
Symbol of solid forms is colour'd light,
And the mute language of the touch is sight. *note*

"Hence in Life's portico starts young Surprise *note*
With step retreating, and expanded eyes;

The virgin, Novelty, whose radiant train
Soars o'er the clouds, or sinks beneath the main,
With sweetly-mutable seductive charms
Thrills the young sense, the tender heart alarms.   150
Then Curiosity with tracing hands
And meeting lips the lines of form demands,   *note*
Buoy'd on light step, o'er ocean, earth, and sky,
Rolls the bright mirror of her restless eye.
While in wild groups tumultuous Passions stand,
And Lust and Hunger head the Motley band;
Then Love and Rage succeed, and Hope and Fear;
And nameless Vices close the gloomy rear;
Or young Philanthropy with voice divine
Convokes the adoring Youth to Virtue's shrine;   160
Who with raised eye and pointing finger leads
To truths celestial, and immortal deeds.

III.

"As the pure language of the Sight commands
The clear ideas furnish'd by the hands;
Beauty's fine forms attract our wondering eyes,
And soft alarms the pausing heart surprise.
Warm from its cell the tender infant born
Feels the cold chill of Life's aerial morn;
Seeks with spread hands the bosoms velvet orbs,   *note*
With closing lips the milky fount absorbs;   170
And, as compress'd the dulcet streams distil,
Drinks warmth and fragrance from the living rill;
Eyes with mute rapture every waving line,
Prints with adoring kiss the Paphian shrine,
And learns erelong, the perfect form confess'd,
Ideal Beauty from its Mother's breast.   *note*

"Now on swift wheels descending like a star

Alights young Eros from his radiant car; *note*
On angel-wings attendant Graces move,
And hail the God of Sentimental Love. 180
Earth at his feet extends her flowery bed, *note*
And bends her silver blossoms round his head;
Dark clouds dissolve, the warring winds subside.
And smiling ocean calms his tossing tide,
O'er the bright morn meridian lustres play,
And Heaven salutes him with a flood of day.

"Warm as the sun-beam, pure as driven snows,
The enamour'd God for young Dione glows;
Drops the still tear, with sweet attention sighs,
And woos the Goddess with adoring eyes; 190
Marks her white neck beneath the gauze's fold,
Her ivory shoulders, and her locks of gold;
Drinks with mute ecstacy the transient glow,
Which warms and tints her bosom's rising snow.
With holy kisses wanders o'er her charms,
And clasps the Beauty in Platonic arms;
Or if the dewy hands of Sleep, unbid,
O'er her blue eye-balls close the lovely lid,
Watches each nascent smile, and fleeting grace,
That plays in day-dreams o'er her blushing face; 200
Counts the fine mazes of the curls, that break
Round her fair ear, and shade her damask cheek;
Drinks the pure fragrance of her breath, and sips
With tenderest touch the roses of her lips;—
O'er female hearts with chaste seduction reigns,
And binds Society in silken chains.

IV.

"If the wide eye the wavy lawns explores, *note*
The bending woodlands, or the winding shores,

Hills, whose green sides with soft protuberance rise,
Or the blue concave of the vaulted skies;—   210
Or scans with nicer gaze the pearly swell
Of spiral volutes round the twisted shell;
Or undulating sweep, whose graceful turns
Bound the smooth surface of Etrurian urns,
When on fine forms the waving lines impress'd
Give the nice curves, which swell the female breast;
The countless joys the tender Mother pours
Round the soft cradle of our infant hours,
In lively trains of unextinct delight
Rise in our bosoms recognized by sight;   220
Fond Fancy's eye recalls the form divine,   *note*
And Taste sits smiling upon Beauty's shrine.

"Where Egypt's pyramids gigantic stand,
And stretch their shadows o'er the shuddering sand;
Or where high rocks o'er ocean's dashing floods
Wave high in air their panoply of woods;
Admiring Taste delights to stray beneath
With eye uplifted, and forgets to breathe;
Or, as aloft his daring footsteps climb,
Crests their high summits with his arm sublime.   *note*

"Where mouldering columns mark the lingering wreck   231
Of Thebes, Palmyra, Babylon, Balbec;
The prostrate obelisk, or shatter'd dome,
Uprooted pedestal, and yawning tomb,
On loitering steps reflective Taste surveys
With folded arms and sympathetic gaze;
Charm'd with poetic Melancholy treads   *note*
O'er ruin'd towns and desolated meads;
Or rides sublime on Time's expanded wings,
And views the fate of ever-changing things.   240

"When Beauty's streaming eyes her woes express,

Or Virtue braves unmerited distress;
Love sighs in sympathy, with pain combined,
And new-born Pity charms the kindred mind;
The enamour'd Sorrow every cheek bedews,
And Taste impassion'd woos the tragic Muse.   *note*

"The rush-thatch'd cottage on the purple moor,
Where ruddy children frolic round the door,
The moss-grown antlers of the aged oak,
The shaggy locks that fringe the colt unbroke,   250
The bearded goat with nimble eyes, that glare
Through the long tissue of his hoary hair;—
As with quick foot he climbs some ruin'd wall,
And crops the ivy, which prevents its fall;—
With rural charms the tranquil mind delight,
And form a picture to the admiring sight.
While Taste with pleasure bends his eye surprised
In modern days at Nature unchastised.   *note*

"The Genius-Form, on silver slippers born,
With fairer dew-drops gems the rising morn;   260
Sheds o'er meridian skies a softer light,
And decks with brighter pearls the brow of night;
With finer blush the vernal blossom glows,
With sweeter breath enamour'd Zephyr blows,
The limpid streams with gentler murmurs pass,
And gayer colours tinge the watery glass,
Charm'd round his steps along the enchanted groves
Flit the fine forms of Beauties, Graces, Loves.

V.

"Alive, each moment of the transient hour,
When Rest accumulates sensorial power,   *note*
The impatient Senses, goaded to contract,   271

Forge new ideas, changing as they act;
And, in long streams dissever'd, or concrete
In countless tribes, the fleeting forms repeat.
Which rise excited in Volition's trains,
Or link the sparkling rings of Fancy's chains;
Or, as they flow from each translucent source,
Pursue Association's endless course.

"Hence when the inquiring hands with contact fine
Trace on hard forms the circumscribing line;   280
Which then the language of the rolling eyes
From distant scenes of earth and heaven supplies;
Those clear ideas of the touch and sight
Rouse the quick sense to anguish or delight;
Whence the fine power of Imitation springs,
And apes the outlines of external things;
With ceaseless action to the world imparts
All moral virtues, languages, and arts.   *note*
First the charm'd Mind mechanic powers collects,
Means for some end, and causes of effects;   290
Then learns from other Minds their joys and fears,
Contagious smiles and sympathetic tears.

"What one fine stimulated Sense discerns,
Another Sense by Imitation learns.—   *note*
So in the graceful dance the step sublime
Learns from the ear the concordance of Time.
So, when the pen of some young artist prints
Recumbent Nymphs in Titian's living tints;
The glowing limb, fair cheek, and flowing hair,
Respiring bosom, and seductive air,   300
He justly copies with enamour'd sigh
From Beauty's image pictured on his eye.

"Thus when great Angelo in wondering Rome   *note*

Fix'd the vast pillars of Saint Peter's dome,
Rear'd rocks on rocks sublime, and hung on high
A new Pantheon in the affrighted sky.
Each massy pier, now join'd and now aloof,
The figured architraves, and vaulted roof,
Ailes, whose broad curves gigantic ribs sustain,
Where holy echoes chant the adoring strain;   310
The central altar, sacred to the Lord,
Admired by Sages, and by Saints ador'd,
Whose brazen canopy ascends sublime
On spiral columns unafraid of Time,
Were first by Fancy in ethereal dyes
Plann'd on the rolling tablets of his eyes;
And his true hand with imitation fine
Traced from his Retina the grand design.

"The Muse of Mimicry in every age   *note*
With silent language charms the attentive stage;   320
The Monarch's stately step, and tragic pause,
The Hero bleeding in his country's cause,
O'er her fond child the dying Mother's tears,
The Lover's ardor, and the Virgin's fears;
The tittering Nymph, that tries her comic task,
Bounds on the scene, and peeps behind her mask,
The Punch and Harlequin, and graver throng,
That shake the theatre with dance and song,
With endless trains of Angers, Loves, and Mirths,
Owe to the Muse of Mimicry their births.   330

"Hence to clear images of form belong
The sculptor's statue, and the poet's song,
The painter's landscape, and the builder's plan,
And Imitation marks the mind of Man.   *note*

VI.

"When strong desires or soft sensations move
The astonish'd Intellect to rage or love;
Associate tribes of fibrous motions rise,
Flush the red cheek, or light the laughing eyes.
Whence ever-active Imitation finds
The ideal trains, that pass in kindred minds; 340
Her mimic arts associate thoughts excite
And the first Language enters at the sight. *note*

"Thus jealous quails or village-cocks inspect
Each other's necks with stiffen'd plumes erect;
Smit with the wordless eloquence, they know
The rival passion of the threatening foe.
So when the famish'd wolves at midnight howl,
Fell serpents hiss, or fierce hyenas growl;
Indignant Lions rear their bristling mail,
And lash their sides with undulating tail. 350
Or when the Savage-Man with clenched fist
Parades, the scowling champion of the list;
With brandish'd arms, and eyes that roll to know
Where first to fix the meditated blow;
Association's mystic power combines
Internal passions with external signs.

"From these dumb gestures first the exchange began
Of viewless thought in bird, and beast, and man;
And still the stage by mimic art displays
Historic pantomime in modern days; 360
And hence the enthusiast orator affords
Force to the feebler eloquence of words.

"Thus the first Language, when we frown'd or smiled,

Rose from the cradle, Imitation's child;
Next to each thought associate sound accords, *note*
And forms the dulcet symphony of words;
The tongue, the lips articulate; the throat *note*
With soft vibration modulates the note;
Love, pity, war, the shout, the song, the prayer
Form quick concussions of elastic air. 370

"Hence the first accents bear in airy rings *note*
The vocal symbols of ideal things,
Name each nice change appulsive powers supply
To the quick sense of touch, or ear or eye.
Or in fine traits abstracted forms suggest
Of Beauty, Wisdom, Number, Motion, Rest;
Or, as within reflex ideas move,
Trace the light steps of Reason, Rage, or Love.
The next new sounds adjunctive thoughts recite,
As hard, odorous, tuneful, sweet, or white. 380
The next the fleeting images select
Of action, suffering, causes and effect;
Or mark existence, with the march sublime
O'er earth and ocean of recording Time.

"The Giant Form on Nature's centre stands,
And waves in ether his unnumber'd hands;
Whirls the bright planets in their silver spheres,
And the vast sun round other systems steers;
Till the last trump amid the thunder's roar
Sound the dread Sentence "Time shall be no more!" 390

"Last steps Abbreviation, bold and strong,
And leads the volant trains of words along;
With sweet loquacity to Hermes springs,
And decks his forehead and his feet with wings.

VII.

"As the soft lips and pliant tongue are taught
With other minds to interchange the thought;
And sound, the symbol of the sense, explains
In parted links the long ideal trains;   *note*
From clear conceptions of external things
The facile power of Recollection springs.   400

"Whence Reason's empire o'er the world presides,   *note*
And man from brute, and man from man divides;
Compares and measures by imagined lines
Ellipses, circles, tangents, angles, sines;
Repeats with nice libration, and decrees
In what each differs, and in what agrees;
With quick Volitions unfatigued selects
Means for some end, and causes of effects;
All human science worth the name imparts,
And builds on Nature's base the works of Arts.   410

"The Wasp, fine architect, surrounds his domes   *note*
With paper-foliage, and suspends his combs;
Secured from frost the Bee industrious dwells,
And fills for winter all her waxen cells;
The cunning Spider with adhesive line
Weaves his firm net immeasurably fine;
The Wren, when embryon eggs her cares engross,
Seeks the soft down, and lines the cradling moss;
Conscious of change the Silkworm-Nymphs begin
Attach'd to leaves their gluten-threads to spin;   420
Then round and round they weave with circling heads
Sphere within Sphere, and form their silken beds.
—Say, did these fine volitions first commence
From clear ideas of the tangent sense;

From sires to sons by imitation caught,
Or in dumb language by tradition taught?
Or did they rise in some primeval site
Of larva-gnat, or microscopic mite;
And with instructive foresight still await
On each vicissitude of insect-state?—   430
Wise to the present, nor to future blind,
They link the reasoning reptile to mankind!
—Stoop, selfish Pride! survey thy kindred forms,
Thy brother Emmets, and thy sister Worms!

"Thy potent acts, Volition, still attend   *note*
The means of pleasure to secure the end;
To express his wishes and his wants design'd
Language, the means, distinguishes Mankind;
For future works in Art's ingenious schools
His hands unwearied form and finish tools;   440
He toils for money future bliss to share,
And shouts to Heaven his mercenary prayer.
Sweet Hope delights him, frowning Fear alarms,
And Vice and Virtue court him to their arms.

"Unenvied eminence, in Nature's plan
Rise the reflective faculties of Man!
Labour to Rest the thinking Few prefer!
Know but to mourn! and reason but to err!—
In Eden's groves, the cradle of the world,
Bloom'd a fair tree with mystic flowers unfurl'd;   450
On bending branches, as aloft it sprung,
Forbid to taste, the fruit of Knowledge hung;
Flow'd with sweet Innocence the tranquil hours,
And Love and Beauty warm'd the blissful bowers.
Till our deluded Parents pluck'd, erelong,
The tempting fruit, and gather'd Right and Wrong;   *note*
Whence Good and Evil, as in trains they pass,
Reflection imaged on her polish'd glass;

And Conscience felt, for blood by Hunger spilt,
The pains of shame, of sympathy, and guilt! 460

VIII.

"Last, as observant Imitation stands,
Turns her quick glance, and brandishes her hands,
With mimic acts associate thoughts excites,
And storms the soul with sorrows or delights;
Life's shadowy scenes are brighten'd and refin'd,
And soft emotions mark the feeling mind. *note*

"The Seraph, Sympathy, from Heaven descends,
And bright o'er earth his beamy forehead bends;
On Man's cold heart celestial ardor flings,
And showers affection from his sparkling wings; 470
Rolls o'er the world his mild benignant eye,
Hears the lone murmur, drinks the whisper'd sigh;
Lifts the closed latch of pale Misfortune's door,
Opes the clench'd hand of Avarice to the poor,
Unbars the prison, liberates the slave,
Sheds his soft sorrows o'er the untimely grave,
Points with uplifted hand to realms above,
And charms the world with universal love.

"O'er the thrill'd frame his words assuasive steal,
And teach the selfish heart what others feel; 480
With sacred truth each erring thought control,
Bind sex to sex, and mingle soul with soul;
From heaven, He cried, descends the moral plan,
And gives Society to savage man.

"High on yon scroll, inscribed o'er Nature's shrine, *note*
Live in bright characters the words divine.
"In Life's disastrous scenes to others do,

What you would wish by others done to you."
—Winds! wide o'er earth the sacred law convey,
Ye Nations, hear it! and ye Kings, obey! 490

"Unbreathing wonder hush'd the adoring throng,
Froze the broad eye, and chain'd the silent tongue;
Mute was the wail of Want, and Misery's cry,
And grateful Pity wiped her lucid eye;
Peace with sweet voice the Seraph-form address'd,
And Virtue clasp'd him to her throbbing breast."

End of Canto III.

NOTES TO CANTO III.

How Oxygen ... (line 13). The atmosphere which surrounds us, is composed of twenty-seven parts of oxygen gas and seventy-three of azote or nitrogen gas, which are simply diffused together, but which, when combined, become nitrous acid. Water consists of eighty-six parts oxygen, and fourteen parts of hydrogen or inflammable air, in a state of combination. It is also probable, that much oxygen enters the composition of glass; as those materials which promote vitrification, contain so much of it, as minium and manganese; and that glass is hence a solid acid in the temperature of our atmosphere, as water is a fluid one.

Two electric streams ... (line 21). It is the opinion of some philosophers, that the electric ether consists of two kinds of fluids diffused together or combined; which are commonly known by the terms of positive and negative electricity, but are by these electricians called vitreous and resinous electricity. The electric shocks given by the torpedo and by the gymnotus, are supposed to be similar to those of the Galvanic pile, as they are produced in water. Which water is decomposed by the Galvanic pile and converted into oxygen and hydrogen gas; see Additional Note XII.

The magnetic ether may also be supposed to consist of two fluids, one of which attracts the needle, and the other repels it; and, perhaps, chemical affinities, and gravitation itself, may consist of two kinds of ether surrounding the particles of bodies, and may thence attract at one distance and repel at another; as appears when two insulated electrised balls are approached to each other, or when two small globules of mercury are pressed together.

And Irritation moves ... (line 64). Irritation is an exertion or change of some extreme part of the sensorium residing in the muscles or organs of sense in consequence of the appulses of external bodies. The word perception includes both the action of the organ of sense in consequence of the impact of external objects and our attention to that action; that is, it expresses both the motion of the organ of sense, or idea, and the pain or pleasure that succeeds or accompanies it. Irritative ideas are those which are preceded by irritation, which is excited by objects external to the organs of sense: as the idea of that tree, which either I attend to, or which I shun in walking near it without attention. In the former case it is termed perception, in the latter it is termed simply an irritative idea.

And young Sensation ... (line 72). Sensation is an exertion or change of the central parts of the sensorium or of the whole of it, beginning at some of those extreme parts of it which reside in the muscles or organs of sense. Sensitive ideas are those which are preceded by the sensation of pleasure or pain, are termed Imagination, and constitute our dreams and reveries.

Quick Volition springs ... (line 73). Volition is an exertion or change of the central parts of the sensorium, or of the whole of it terminating in some of those extreme parts of it which reside in the muscles and organs of sense. The vulgar use of the word memory is too unlimited for our purpose: those ideas which we voluntarily recall are here termed ideas of recollection, as when we will to repeat the alphabet backwards. And those ideas which are suggested to us by preceding ideas are here termed ideas of suggestion, as whilst we repeat the alphabet in the usual order; when by habits previously acquired B is suggested by A, and C by B, without any effort of deliberation. Reasoning is that operation of the sensorium by which we excite two or many tribes of ideas, and then reexcite the ideas in which they differ or correspond. If we determine this difference, it is called judgment; if we in vain endeavour to determine it, it is called doubting.

If we reexcite the ideas in which they differ, it is called distinguishing. If we reexcite those in which they correspond, it is called comparing.

Each passing moment ... (line 81). During our waking hours, we perpetually compare the passing trains of our ideas with the known system of nature, and reject those which are incongruous with it; this is explained in Zoonomia, Sect. XVII, 3, 7, and is there termed Intuitive Analogy. When we sleep, the faculty of volition ceases to act, and in consequence the uncompared trains of ideas become incongruous and form the farrago of our dreams; in which we never experience any surprise, or sense of novelty.

Association steers ... (line 91). Association is an exertion or change of some extreme part of the sensorium residing in the muscles and organs of sense in consequence of some antecedent or attendant fibrous contractions. Associate ideas, therefore, are those which are preceded by other ideas or muscular motions, without the intervention of irritation, sensation, or volition between them; these are also termed ideas of suggestion.

The branching forehead ... (line 103). The peculiarities of the shapes of animals which distinguish them from each other, are enumerated in Zoonomia, Sect. XXXIX, 4, 8, on Generation, and are believed to have been gradually formed from similar living fibres, and are varied by reproduction. Many of these parts of animals are there shown to have arisen from their three great desires of lust, hunger, and security.

The tropic eel ... (line 111). Gymnotus electricus.

The fly of night ... (line 113). Lampyris noctiluca. Fire-fly.

The hand, first gift of Heaven ... (line 122). The human species in some of their sensations are much inferior to animals, yet the accuracy of the sense of touch, which they possess in so eminent a degree, gives them a great superiority of understanding; as is well observed by the ingenious Mr. Buffon. The extremities of other animals terminate in horns, and hoofs, and claws, very unfit for the sensation of touch; whilst the human hand is finely adapted to encompass its object with this organ of sense. Those animals who have clavicles or collar-bones, and thence use their forefeet like hands, as cats, squirrels, monkeys, are more ingenious than other quadrupeds, except the elephant, who has a fine sense at the extremity of his proboscis; and many insects from the possessing finer organs of touch have greater ingenuity, as spiders, bees, wasps.

Trace the nice lines of form ... (line 125). When the idea of solidity is excited a part of the extensive organ of touch is compressed by some external body, and this part of the sensorium so compressed exactly resembles in figure the figure of the body that compressed it. Hence when we acquire the idea of solidity, we acquire at the same time the idea of figure; and this idea of figure, or motion of a part of the organ of touch, exactly resembles in its figure the figure of the body that occasions it; and thus exactly acquaints us with this property of the external world.

Now, as the whole universe with all its parts possesses a certain form or figure, if any part of it moves, that form or figure of the whole is varied. Hence, as motion is no other than a perpetual variation of figure, our idea of motion is also a real resemblance of the motion that produced it.

Hence arises the certainty of the mathematical sciences, as they explain these properties of bodies, which are exactly resembled by our ideas of them, whilst we are obliged to collect almost all our other knowledge from experiment; that is, by observing the effects exerted by one body upon another.

The mute language of the touch ... (line 144). Our eyes observe a difference of colour, or of shade, in the prominences and depressions of objects, and that those shades uniformly vary when the sense of touch observes any variation. Hence when the retina becomes stimulated by colours or shades of light in a certain form, as in a circular spot, we know by experience that this is a sign that a tangible body is before us; and that its figure is resembled by the miniature figure of the part of the organ of vision that is thus stimulated.

Here whilst the stimulated part of the retina resembles exactly the visible figure of the whole in miniature, the various kinds of stimuli from different colours mark the visible figures of the minuter parts; and by habit we instantly recall the tangible figures.

So that though our visible ideas resemble in miniature the outline of the figure of coloured bodies, in other respects they serve only as a language, which by acquired associations introduce the tangible ideas of bodies. Hence it is, that this sense is so readily deceived by the art of the painter to our amusement and instruction. The reader will find much very curious knowledge on this subject in Bishop Berkeley's Essay on Vision, a work of great ingenuity.

Starts young Surprise ... (line 145). Surprise is occasioned by the sudden interruption of the usual trains of our ideas by any violent stimulus from external objects, as from the unexpected discharge of a pistol, and hence does not exist in our dreams, because our external senses are closed or inirritable. The fetus in the womb must experience many sensations, as of resistance, figure, fluidity, warmth, motion, rest, exertion, taste; and must consequently possess trains both of waking and sleeping ideas. Surprise must therefore be strongly excited at its nativity, as those trains of ideas must instantly be dissevered by the sudden and violent sensations occasioned by the dry and cold atmosphere, the hardness of external bodies, light, sound, and odours; which are accompanied with pleasure or pain according to their quantity or intensity.

As some of these sensations become familiar by repetition, other objects not previously attended to present themselves, and produce the idea of novelty, which is a less degree of surprise, and like that is not perceived in our dreams, though for another reason; because in sleep we possess no voluntary power to compare our trains of ideas with our previous knowledge of nature, and do not therefore perceive their difference by intuitive analogy from what usually occurs.

As the novelty of our ideas is generally attended with pleasurable sensation, from this arises Curiosity, or a desire of examining a variety of objects, hoping to find novelty, and the pleasure consequent to this degree of surprise; see Additional Note VII, 3.

And meeting lips ... (line 152). Young children put small bodies into their mouths, when they are satiated with food, as well as when they are hungry, not with design to taste them, but use their lips as an organ of touch to distinguish the shape of them. Puppies, whose toes are terminated with nails, and who do not much use their forefeet as hands, seem to have no other means of acquiring a knowledge of the forms of external bodies, and are therefore perpetually playing with things by taking them between their lips.

Seeks with spread hands ... (line 169). These eight beautiful lines are copied from Mr. Bilsborrow's Address prefixed to Zoonomia, and are translated from that work; Sect. XVI, 6.

Ideal Beauty ... (line 176). Sentimental Love, as distinguished from the animal passion of that name, with which it is frequently accompanied, consists in the desire or sensation of beholding, embracing, and saluting a beautiful object.

The characteristic of beauty therefore is that it is the object of love; and though many other objects are in common language called beautiful, yet they are only called so metaphorically, and ought to be termed agreeable. A Grecian temple may give us the pleasurable idea of sublimity, a Gothic temple may give us the pleasurable idea of variety, and a modern house the pleasurable idea of utility; music and poetry may inspire our love by association of ideas; but none of these, except metaphorically, can be termed beautiful, as we have no wish to embrace or salute them.

Our perception of beauty consists in our recognition by the sense of vision of those objects, first, which have before inspired our love by the pleasure, which they have afforded to many of our senses; as to our sense of warmth, of touch, of smell, of taste, hunger and thirst; and, secondly, which bear any analogy of form to such objects.

Alights young Eros ... (line 178). There were two deities of Love belonging to the heathen mythology, the one said to be celestial, and the other terrestrial. Aristophanes says, "Sable-winged Night produced an egg, from which sprung up like a blossom Eros, the lovely, the desirable, with his glossy golden wings." See Botanic Garden, Canto I, line 412, Note. The other deity of Love, Cupido, seems of much later date, as he is not mentioned in the works of Homer, where there were so many apt situations to have introduced him.

Earth at his feet ... (line 181).
Te, Dea, te fugiunt venti, te nubila coeli,
Adventumque tuum; tibi suaves daedala tellus
Submittit flores; tibi rident aequora ponti;
Placatumque nitet diffuso lumine coelum.
[Before thee, Goddess, and thy coming on / Flee stormy wind and massy cloud away / For thee the daedal Earth bears scented flowers / For thee waters of the unvexed deep / Smile, and the hollows of the serene sky / Glow with diffused radiance for thee!]
              Lucretius.

The wavy lawns ... (line 207). When the babe, soon after it is born into this cold world, is applied to its mother's bosom; its sense of perceiving warmth is first agreeably affected; next its sense of smell is delighted with the odour of her milk; then its taste is gratified by the flavour of it; afterwards the appetites of hunger and of thirst afford pleasure by the possession of their objects, and by the subsequent digestion of the aliment; and lastly, the sense of touch is delighted by the softness and smoothness of the milky fountain, the source of such variety of happiness.

All these various kinds of pleasure at length become associated with the form of the mother's breast; which the infant embraces with its hands, presses with its lips, and watches with its eyes; and thus acquires more accurate ideas of the form of its mother's bosom, than of the odour and flavour or warmth, which it perceives by its other senses. And hence at our maturer years, when any object of vision is presented to us, which by its waving or spiral lines bears any similitude to the form of the female bosom, whether it be found in a landscape with soft gradations of rising and descending surface, or in the forms of some antique vases, or in other works of the pencil or the chisel, we feel a general glow of delight, which seems to influence all our senses; and if the object be not too large, we experience an attraction to embrace it with our arms, and to salute it with our lips, as we did in our early infancy the bosom of our mother. And thus we find, according to the ingenious idea of Hogarth, that the waving lines of beauty were originally taken from the temple of Venus.

Fond Fancy's eye recalls ... (line 221). See Additional Note XIII.

With his arm sublime ... (line 230). Objects of taste have been generally divided into the beautiful, the sublime, and the new; and lately to these have been added the picturesque. The beautiful so well explained in Hogarth's analysis of beauty, consists of curved lines and smooth surfaces, as expressed in the preceding note; any object larger than usual, as a very large temple or a very large mountain, gives us the idea of sublimity; with which is often confounded the terrific, and the melancholic: what is now termed picturesque includes objects, which are principally neither sublime nor beautiful, but which by their variety and intricacy joined with a due degree of regularity or uniformity convey to the mind an agreeable sentiment of novelty. Many other agreeable sentiments may be excited by visible objects, thus to the sublime and beautiful may be added the terrific, tragic, melancholic, artless, &c. while novelty superinduces a charm upon them all. See Additional Note XIII.

Poetic melancholy treads ... (line 237). The pleasure arising from the contemplation of the ruins of ancient grandeur or of ancient happiness, and here termed poetic melancholy, arises from a combination of the painful idea of sorrow with the pleasurable idea of the grandeur or happiness of past times; and becomes very interesting to us by fixing our attention more strongly on that grandeur and happiness, as the passion of Pity mentioned in the succeeding note is a combination of the painful idea of sorrow with the pleasurable one of beauty, or of virtue.

The tragic Muse ... (line 246). Why we are delighted with the scenical representations of Tragedy, which draw tears from our eyes, has been variously explained by different writers. The same distressful circumstance attending an ugly or wicked person affects us with grief or disgust; but when distress occurs to a beauteous or virtuous person, the pleasurable idea of beauty or of virtue becomes mixed with the painful one of sorrow and the passion of Pity is produced, which is a combination of love or esteem with sorrow; and becomes highly interesting to us by fixing our attention more intensely on the beauteous or virtuous person.

Other distressful scenes have been supposed to give pleasure to the spectator from exciting a comparative idea of his own happiness, as when a shipwreck is viewed by a person safe on shore, as mentioned by Lucretius, line 3. But these dreadful situations belong rather to the terrible, or the horrid, than to the tragic; and may be objects of curiosity from their novelty, but not of Taste, and must suggest much more pain than pleasure.

Nature unchastised ... (line 258). In cities or their vicinity, and even in the cultivated parts of the country we rarely see undisguised nature; the fields are ploughed, the meadows mown, the shrubs planted in rows for hedges, the trees deprived of their lower branches, and the animals, as horses, dogs, and sheep, are mutilated in respect to their tails or ears; such is the useful or ill-employed activity of mankind! all which alterations add to the formality of the soil, plants, trees, or animals; whence when natural objects are occasionally presented to us, as an uncultivated forest and its wild inhabitants, we are not only amused with greater variety of form, but are at the same time enchanted by the charm of novelty, which is a less degree of Surprise, already spoken of in note on line 145 of this Canto.

When rest accumulates ... (line 270). The accumulation of the spirit of animation, when those parts of the system rest, which are usually in motion, produces a disagreeable sensation. Whence the pain of cold and of hunger, and the irksomeness of a continued attitude, and of an indolent life: and hence the propensity to action in those confined animals, which have been accustomed to activity, as is seen in the motions of a squirrel in a cage; which uses perpetual exertion to exhaust a part of its accumulated sensorial power. This is one source of our general propensity to action; another perhaps arises from our curiosity or expectation of novelty mentioned in the note on line 145 of this Canto.

But the immediate cause of our propensity to imitation above that of other animals arises from the greater facility, with which by the sense of touch we acquire the ideas of the outlines of objects, and afterwards in consequence by the sense of sight; this seems to have been observed by Aristotle, who calls man, "the imitative animal;" see Zoonomia, Vol. I, Sect. XXII.

All moral virtues ... (line 288). See the sequel of this canto line 453 on sympathy; and line 331 on language; and the subsequent lines on the arts of painting and architecture.

Another sense ... (line 294). As the part of the organs of touch or of sight, which is stimulated into action by a tangible or visible object, must resemble in figure at least the figure of that object, as it thus constitutes an idea; it may be said to imitate the figure of that object; and thus imitation may be esteemed coeval with the existence both of man and other animals: but this would confound perception with imitation; which latter is better defined from the actions of one sense copying those of another.

Thus when great Angelo ... (line 303). The origin of this propensity to imitation has not been deduced from any known principle; when any action presents itself to the view of a child, as of whetting a knife, or threading a needle; the parts of this action in respect of time, motion, figure, are imitated by parts of the retina of his eye; to perform this action therefore with his hands is easier to him than to invent any new action; because it consists in repeating with another set of fibres, viz. with the moving muscles, what he had just performed by some parts of the retina; just as in dancing we transfer the times of the motions from the actions of the auditory nerves to the muscles of the limbs. Imitation therefore consists of repetition, which is the easiest kind of animal action; as the ideas or motions become presently associated together;

which adds to the facility of their production; as shown in Zoonomia, Vol. I, Sect. XXII, 2.

It should be added, that as our ideas, when we perceive external objects, are believed to consist in the actions of the immediate organs of sense in consequence of the stimulus of those objects; so when we think of external objects, our ideas are believed to consist in the repetitions of the actions of the immediate organs of sense, excited by the other sensorial powers of volition, sensation, or association.

The Muse of Mimicry ... (line 319). Much of the pleasure received from the drawings of flowers finely finished, or of portraits, is derived from their imitation or resemblance of the objects or persons which they represent. The same occurs in the pleasure we receive from mimicry on the stage; we are surprised at the accuracy of its enacted resemblance. Some part of the pleasure received from architecture, as when we contemplate the internal structure of gothic temples, as of King's College chapel in Cambridge, or of Lincoln Cathedral, may arise also from their imitation or resemblance of those superb avenues of large trees, which were formerly appropriated to religious ceremonies.

Imitation marks ... (line 334). Many other curious instances of one part of the animal system imitating another part of it, as in some contagious diseases; and also of some animals imitating each other, are given in Zoonomia, Vol. I, Sect. XXII, 3. To which may be added, that this propensity to imitation not only appears in the actions of children, but in all the customs and fashions of the world; many thousands tread in the beaten paths of others, who precede or accompany them, for one who traverses regions of his own discovery.

And the first Language ... (line 342). There are two ways by which we become acquainted with the passions of others: first, by having observed the effects of them, as of fear or anger, on our own bodies, we know at sight when others are under the influence of these affections. So children long before they can speak, or understand the language of their parents, may be frightened by an angry countenance, or soothed by smiles and blandishments.

Secondly, when we put ourselves into the attitude that any passion naturally occasions, we soon in some degree acquire that passion; hence when those that scold indulge themselves in loud oaths and violent actions of the arms, they increase their anger by the mode of expressing themselves; and, on the contrary, the counterfeited smile of pleasure in disagreeable company soon brings along with it a portion of the reality, as is well illustrated by Mr. Burke. (Essay on the Sublime and Beautiful.)

These are natural signs by which we understand each other, and on this slender basis is built all human language. For without some natural signs no artificial ones could have been invented or understood, as is very ingeniously observed by Dr. Reid. (Inquiry into the Human Mind.)

Next to each thought ... (line 365). See Additional Note XIV.

The tongue, the lips articulate ... (line 367). See Additional Note XV.

Hence the first accents ... (line 371). Words were originally the signs or names of individual ideas; but in all known languages many of them by changing their terminations express more than one idea, as in the cases of nouns, and the moods and tenses of verbs. Thus a whip suggests a single idea of that instrument; but "to whip," suggests an idea of action, joined with that of the instrument, and is then called a verb; and "to be whipped," suggests an idea of being acted upon or suffering. Thus in most languages two ideas are suggested by one word by changing its termination; as amor, love; amare, to love; amari, to be loved.

Nouns are the names of the ideas of things, first as they are received by the stimulus of objects, or as they are afterwards repeated; secondly, they are names of more abstracted ideas, which do not suggest at the same time the external objects, by which they were originally excited; or thirdly, of the operations of our minds, which are termed reflex ideas by metaphysical writers; or lastly, they are the names of our ideas of parts or properties of objects; and are termed by grammarians nouns adjective.

Verbs are also in reality names of our ideas of things, or nouns, with the addition of another idea to them, as of acting or suffering; or of more than one other annexed idea, as of time, and also of existence. These with the numerous abbreviations, so well illustrated by Mr. Horne Tooke in his Diversions of Purley, make up the general theory of language, which consists of the symbols of ideas represented by vocal or written words; or by parts of those words, as their terminations; or by their disposition in respect to their order or succession; as further explained in Additional Note XIV.

In parted links ... (line 398). As our ideas consist of successive trains of the motions, or changes of figure, of the extremities of the nerves of one or more of our senses, as of the optic or auditory nerves; these successive trains of motion, or configuration, are in common life divided into many links, to each of which a word or name is given, and it is called an idea. This chain of ideas may be broken into more or fewer links, or divided in different parts of it, by the customs of different people. Whence the meanings of the words of one language cannot always be exactly expressed by those of another; and hence the acquirement of different languages in their infancy may affect the modes of thinking and reasoning of whole nations, or of different classes of society; as the words of them do not accurately suggest the same ideas, or parts of ideal trains; a circumstance which has not been sufficiently analysed.

Whence Reason's empire ... (line 401). The facility of the use of the voluntary power, which is owing to the possession of the clear ideas acquired by our superior sense of touch, and afterwards of vision, distinguishes man from brutes, and has given him the empire of the world, with the power of improving nature by the exertions of art.

Reasoning is that operation of the sensorium by which we excite two or many tribes of ideas, and then reexcite the ideas in which they differ or correspond. If we determine this difference, it is called judgment; if we in vain endeavour to determine it, it is called doubting.

If we reexcite the ideas in which they differ, it is called distinguishing. If we reexcite those in which they correspond, it is called comparing.

The Wasp, fine architect ... (line 411). Those animals which possess a better sense of touch are, in general, more ingenious than others. Those which have claviculae, or collar-bones, and thence use the forefeet as hands, as the monkey, squirrel, rat, are more ingenious in seizing their prey or escaping from danger. And the ingenuity of the elephant appears to arise from the sense of touch at the extremity of his proboscis, which has a prominence on one side of its cavity like a thumb to close against the other side of it, by which I have seen him readily pick up a shilling which was thrown amongst the straw he stood upon. Hence the excellence of the sense of touch in many insects seems to have given them wonderful ingenuity so as to equal or even excel mankind in some of their arts and discoveries; many of which may have been acquired in situations previous to their present ones, as the great globe itself, and all that it inhabit, appear to be in a perpetual state of mutation and improvement; see Additional Note IX.

Thy potent acts, Volition ... (line 435). It was before observed, how much the superior accuracy of our sense of touch contributes to increase our knowledge; but it is the greater energy and activity of the power of volition, that marks mankind, and has given them the empire of the world.

There is a criterion by which we may distinguish our voluntary acts or thoughts from those that are excited by our sensations: "The former are always employed about the means to acquire pleasurable objects, or to avoid painful ones; while the latter are employed about the possession of those that are already in our power."

The ideas and actions of brutes, like those of children, are almost perpetually produced by their present pleasures or their present pains; and they seldom busy themselves about the means of procuring future bliss, or of avoiding future misery.

Whilst the acquiring of languages, the making of tools, and the labouring for money, which are all only the means of procuring pleasure; and the praying to the Deity, as another means to procure happiness, are characteristic of human nature.

And gather'd Right and Wrong ... (line 456). Some philosophers have believed that the acquisition of knowledge diminishes the happiness of the possessor; an opinion which seems to have been inculcated by the history of our first parents, who are said to have become miserable from eating of the tree of knowledge. But as the foresight and the power of mankind are much increased by their voluntary exertions in the acquirement of knowledge, they may undoubtedly avoid many sources of evil, and procure many sources of good; and yet possess the pleasures of sense, or of imagination, as extensively as the brute or the savage.

And soft emotions ... (line 466). From our aptitude to imitation arises what is generally understood by the word sympathy, so well explained by Dr. Smith of Glasgow. Thus the appearance of a cheerful countenance gives us pleasure, and of a melancholy one makes us sorrowful. Yawning, and sometimes vomiting, are thus propagated by sympathy; and some people of delicate fibres, at the presence of a spectacle of misery, have felt pain in the same parts of their bodies, that were diseased or mangled in the object they saw.

The effect of this powerful agent in the moral world, is the foundation of all our intellectual sympathies with the pains and pleasures of others, and is in consequence the source of all our virtues. For in what consists our sympathy with the miseries or with the joys of our fellow creatures, but in an involuntary excitation of ideas in some measure similar or imitative of those which we believe to exist in the minds of the persons whom we commiserate or congratulate!

High on yon scroll ... (line 485). The famous sentence of Socrates "Know thyself," so celebrated by writers of antiquity, and said by them to have descended from Heaven, however wise it may be, seems to be rather of a selfish nature; and the author of it might have added "Know also other people." But the sacred maxims of the author of Christianity, "Do as you would be done by," and "Love your neighbour as yourself," include all our duties of benevolence and morality; and, if sincerely obeyed by all nations, would a thousandfold multiply the present happiness of mankind.

Canto IV.
OF GOOD AND EVIL

I.

"How few," the Muse in plaintive accents cries,
And mingles with her words pathetic sighs.—
"How few, alas! in Nature's wide domains
The sacred charm of Sympathy restrains!
Uncheck'd desires from appetite commence,
And pure reflection yields to selfish sense!
—Blest is the Sage, who learn'd in Nature's laws   *note*
With nice distinction marks effect and cause;
Who views the insatiate Grave with eye sedate,
Nor fears thy voice, inexorable Fate!   10

"When War, the Demon, lifts his banner high,
And loud artillery rends the affrighted sky;
Swords clash with swords, on horses horses rush,
Man tramples man, and nations nations crush;
Death his vast sithe with sweep enormous wields,
And shuddering Pity quits the sanguine fields.

"The wolf, escorted by his milk-drawn dam,
Unknown to mercy, tears the guiltless lamb;
The towering eagle, darting from above,   *note*
Unfeeling rends the inoffensive dove;   20
The lamb and dove on living nature feed,
Crop the young herb, or crush the embryon seed.

Nor spares the loud owl in her dusky flight,
Smit with sweet notes, the minstrel of the night;
Nor spares, enamour'd of his radiant form,
The hungry nightingale the glowing worm;
Who with bright lamp alarms the midnight hour,
Climbs the green stem, and slays the sleeping flower.

"Fell Oestrus buries in her rapid course   *note*
Her countless brood in stag, or bull, or horse;   30
Whose hungry larva eats its living way,
Hatch'd by the warmth, and issues into day.
The wing'd Ichneumon for her embryon young   *note*
Gores with sharp horn the caterpillar throng.
The cruel larva mines its silky course,
And tears the vitals of its fostering nurse.
While fierce Libellula with jaws of steel   *note*
Ingulfs an insect-province at a meal;
Contending bee-swarms rise on rustling wings,   *note*
And slay their thousands with envenom'd stings.   40

"Yes! smiling Flora drives her armed car
Through the thick ranks of vegetable war;
Herb, shrub, and tree, with strong emotions rise
For light and air, and battle in the skies;
Whose roots diverging with opposing toil
Contend below for moisture and for soil;
Round the tall Elm the flattering Ivies bend,
And strangle, as they clasp, their struggling friend;
Envenom'd dews from Mancinella flow,
And scald with caustic touch the tribes below;   50
Dense shadowy leaves on stems aspiring borne
With blight and mildew thin the realms of corn;
And insect hordes with restless tooth devour
The unfolded bud, and pierce the ravell'd flower.

"In ocean's pearly haunts, the waves beneath
Sits the grim monarch of insatiate Death;
The shark rapacious with descending blow			*note*
Darts on the scaly brood, that swims below;
The crawling crocodiles, beneath that move,			*note*
Arrest with rising jaw the tribes above;			60
With monstrous gape sepulchral whales devour
Shoals at a gulp, a million in an hour.
—Air, earth, and ocean, to astonish'd day
One scene of blood, one mighty tomb display!
From Hunger's arm the shafts of Death are hurl'd,
And one great Slaughter-house the warring world!			*note*

"The brow of Man erect, with thought elate,
Ducks to the mandate of resistless fate;
Nor Love retains him, nor can Virtue save
Her sages, saints, or heroes from the grave.			70
While cold and hunger by defect oppress,			*note*
Repletion, heat, and labour by excess,
The whip, the sting, the spur, the fiery brand,
And, cursed Slavery! thy iron hand;
And led by Luxury Disease's trains,
Load human life with unextinguish'd pains.

"Here laughs Ebriety more fell than arms,			*note*
And thins the nations with her fatal charms,
With Gout, and Hydrops groaning in her train,
And cold Debility, and grinning Pain,			80
With harlot's smiles deluded man salutes,
Revenging all his cruelties to brutes!
There the curst spells of Superstition blind,
And fix her fetters on the tortured mind;
She bids in dreams tormenting shapes appear,
With shrieks that shock Imagination's ear,
E'en o'er the grave a deeper shadow flings,			*note*

And maddening Conscience darts a thousand stings.

"There writhing Mania sits on Reason's throne,
Or Melancholy marks it for her own,   90
Sheds o'er the scene a voluntary gloom,
Requests oblivion, and demands the tomb.
And last Association's trains suggest   *note*
Ideal ills, that harrow up the breast,   *note*
Call for the dead from Time's o'erwhelming main,
And bid departed Sorrow live again.

"Here ragged Avarice guards with bolted door
His useless treasures from the starving poor;
Loads the lorn hours with misery and care,
And lives a beggar to enrich his heir.   *note*
Unthinking crowds thy forms, Imposture, gull,   101
A Saint in sackcloth, or a Wolf in wool.   *note*
While mad with foolish fame, or drunk with power,
Ambition slays his thousands in an hour;
Demoniac Envy scowls with haggard mien,
And blights the bloom of other's joys, unseen;
Or wrathful Jealousy invades the grove,
And turns to night meridian beams of Love!

"Here wide o'er earth impetuous waters sweep,
And fields and forests rush into the deep;   110
Or dread Volcano with explosion dire
Involves the mountains in a flood of fire;
Or yawning Earth with closing jaws inhumes
Unwarned nations, living in their tombs;
Or Famine seizes with her tiger-paw,
And swallows millions with unsated maw.

"There livid Pestilence in league with Dearth
Walks forth malignant o'er the shuddering earth,
Her rapid shafts with airs volcanic wings,   *note*

Or steeps in putrid vaults her venom'd stings.　　　120
Arrests the young in Beauty's vernal bloom,
And bears the innocuous strangers to the tomb!—

"And now, e'en I, whose verse reluctant sings
The changeful state of sublunary things,
Bend o'er Mortality with silent sighs,
And wipe the secret tear-drops from my eyes,
Hear through the night one universal groan,
And mourn unseen for evils not my own,
With restless limbs and throbbing heart complain,
Stretch'd on the rack of sentimental pain!　　　*note*
—Ah where can Sympathy reflecting find　　　131
One bright idea to console the mind?
One ray of light in this terrene abode
To prove to Man the Goodness of his God?"

II.

"Hear, O ye Sons of Time!" the Nymph replies,
Quick indignation darting from her eyes;
"When in soft tones the Muse lamenting sings,
And weighs with tremulous hand the sum of things;
She loads the scale in melancholy mood,
Presents the evil, but forgets the good.　　　140
But if the beam some firmer hand suspends,
And good and evil load the adverse ends;
With strong libration, where the Good abides,
Quick nods the beam, the ponderous gold subsides.

"Hear, O ye Sons of Time! the powers of Life
Arrest the elements, and stay their strife;
From wandering atoms, ethers, airs, and gas,　　　*note*
By combination form the organic mass;
And,—as they seize, digest, secrete,—dispense

The bliss of Being to the vital Ens.   150
Hence in bright groups from Irritation rise
Young Pleasure's trains, and roll their azure eyes.

"With fond delight we feel the potent charm,
When Zephyrs cool us, or when sun-beams warm;
With fond delight inhale the fragrant flowers,
Taste the sweet fruits, which bend the blushing bowers,
Admire the music of the vernal grove,
Or drink the raptures of delirious love.

"So with long gaze admiring eyes behold
The varied landscape all its lights unfold;   *note*
Huge rocks opposing o'er the stream project   161
Their naked bosoms, and the beams reflect;
Wave high in air their fringed crests of wood,
And checker'd shadows dance upon the flood;
Green sloping lawns construct the sidelong scene,
And guide the sparkling rill that winds between;
Conduct on murmuring wings the pausing gale,
And rural echoes talk along the vale;
Dim hills behind in pomp aerial rise,
Lift their blue tops, and melt into the skies.   170

"So when by Handel tuned to measured sounds
The trumpet vibrates, or the drum rebounds;
Alarm'd we listen with ecstatic wonder
To mimic battles, or imagined thunder.
When the soft lute in sweet impassion'd strains
Of cruel nymphs or broken vows complains;
As on the breeze the fine vibration floats,
We drink delighted the melodious notes.   *note*
But when young Beauty on the realms above
Bends her bright eye, and trills the tones of love;   180
Seraphic sounds enchant this nether sphere;

And listening angels lean from Heaven to hear.

"Next by Sensation led, new joys commence
From the fine movements of the excited sense;
In swarms ideal urge their airy flight,
Adorn the day-scenes, and illume the night.
Her spells o'er all the hand of Fancy flings,
Gives form and substance to unreal things;
With fruits and foliage decks the barren waste,
And brightens Life with sentiment and taste; 190
Pleased o'er the level and the rule presides,
The painter's brush, the sculptor's chisel guides,
With ray ethereal lights the poet's fire,
Tunes the rude pipe, or strings the heroic lyre:
Charm'd round the nymph on frolic footsteps move
The angelic forms of Beauty, Grace, and Love.

"So dreams the Patriot, who indignant draws
The sword of vengeance in his Country's cause;
Bright for his brows unfading honours bloom,
Or kneeling Virgins weep around his tomb. 200
So holy transports in the cloister's shade
Play round thy toilet, visionary maid!
Charm'd o'er thy bed celestial voices sing,
And Seraphs hover on enamour'd wing.

"So Howard, Moira, Burdett, sought the cells,
Where want, or woe, or guilt in darkness dwells;
With Pity's torch illumed the dread domains,
Wiped the wet eye, and eased the galling chains;
With Hope's bright blushes warm'd the midnight air,
And drove from earth the Demon of Despair. 210
Erewhile emerging from the caves of night
The Friends of Man ascended into light;
With soft assuasive eloquence address'd

The ear of Power to stay his stern behest;
At Mercy's call to stretch his arm and save
His tottering victims from the gaping grave.
These with sweet smiles Imagination greets,
For these she opens all her treasured sweets,
Strews round their couch, by Pity's hand combined,
Bright flowers of joy, the sunshine of the mind; 220
While Fame's loud trump with sounds applausive breathes
And Virtue crowns them with immortal wreathes.

"Thy acts, Volition, to the world impart
The plans of Science with the works of art;
Give to proud Reason her comparing power,
Warm every clime, and brighten every hour.
In Life's first cradle, ere the dawn began
Of young Society to polish man;
The staff that propp'd him, and the bow that arm'd,
The boat that bore him, and the shed that warm'd, 230
Fire, raiment, food, the ploughshare, and the sword,
Arose, Volition, at thy plastic word.

"By thee instructed, Newton's eye sublime
Mark'd the bright periods of revolving time;
Explored in Nature's scenes the effect and cause,
And, charm'd, unravell'd all her latent laws.
Delighted Herschel with reflected light
Pursues his radiant journey through the night;
Detects new guards, that roll their orbs afar
In lucid ringlets round the Georgian star. 240

"Inspired by thee, with scientific wand
Pleased Archimedes mark'd the figured sand; *note*
Seized with mechanic grasp the approaching decks,
And shook the assailants from the inverted wrecks.
—Then cried the Sage, with grand effects elate,
And proud to save the Syracusian state;

While crowds exulting shout their noisy mirth,
'Give where to stand, and I will move the earth.'
So Savery guided his explosive steam   *note*
In iron cells to raise the balanced beam;   250
The Giant-form its ponderous mass uprears,
Descending nods and seems to shake the spheres.

"Led by Volition on the banks of Nile
Where bloom'd the waving flax on Delta's isle,   *note*
Pleased Isis taught the fibrous stems to bind,
And part with hammers from the adhesive rind;
With locks of flax to deck the distaff-pole,
And whirl with graceful bend the dancing spole.
In level lines the length of woof to spread,
And dart the shuttle through the parting thread.   260
So Arkwright taught from Cotton-pods to cull,   *note*
And stretch in lines the vegetable wool;
With teeth of steel its fibre-knots unfurl'd,
And with the silver tissue clothed the world.

"Ages remote by thee, Volition, taught
Chain'd down in characters the winged thought;
With silent language mark'd the letter'd ground,
And gave to sight the evanescent sound.
Now, happier lot! enlighten'd realms possess
The learned labours of the immortal Press;   *note*
Nursed on whose lap the births of science thrive,   271
And rising Arts the wrecks of Time survive.

"Ye patriot heroes! in the glorious cause
Of Justice, Mercy, Liberty, and Laws,
Who call to Virtue's shrine the British youth,
And shake the senate with the voice of Truth;
Rouse the dull ear, the hoodwink'd eye unbind,
And give to energy the public mind;
While rival realms with blood unsated wage

Wide-wasting war with fell demoniac rage; 280
In every clime while army army meets,
And oceans groan beneath contending fleets;
Oh save, oh save, in this eventful hour
The tree of knowledge from the axe of power;
With fostering peace the suffering nations bless,
And guard the freedom of the immortal Press!
So shall your deathless fame from age to age
Survive recorded in the historic page;
And future bards with voice inspired prolong
Your sacred names immortalized in song. 290

"Thy power Association next affords
Ideal trains annex'd to volant words,
Conveys to listening ears the thought superb,
And gives to Language her expressive verb; *note*
Which in one changeful sound suggests the fact
At once to be, to suffer, or to act;
And marks on rapid wing o'er every clime
The viewless flight of evanescent Time.

"Call'd by thy voice contiguous thoughts embrace *note*
In endless streams arranged by Time or Place; 300
The Muse historic hence in every age
Gives to the world her interesting page;
While in bright landscape from her moving pen
Rise the fine tints of manners and of men.

"Call'd by thy voice Resemblance next describes
Her sister-thoughts in lucid trains or tribes;
Whence pleased Imagination oft combines
By loose analogies her fair designs;
Each winning grace of polish'd wit bestows *note*
To deck the Nymphs of Poetry and Prose. 310

"Last, at thy potent nod, Effect and Cause

Walk hand in hand accordant to thy laws;
Rise at Volition's call, in groups combined,
Amuse, delight, instruct, and serve Mankind;
Bid raised in air the ponderous structure stand,
Or pour obedient rivers through the land;
With cars unnumber'd crowd the living streets,
Or people oceans with triumphant fleets.

"Thy magic touch imagined forms supplies
From colour'd light, the language of the eyes;   320
On Memory's page departed hours inscribes,
Sweet scenes of youth, and Pleasure's vanish'd tribes.
By thee Antinous leads the dance sublime
On wavy step, and moves in measured time;
Charm'd round the Youth successive Graces throng,
And Ease conducts him, as he moves along;
Unbreathing crowds the floating form admire,
And Vestal bosoms feel forbidden fire.

"When rapp'd Cecilia breathes her matin vow,
And lifts to Heaven her fair adoring brow;   330
From her sweet lips, and rising bosom part
Impassion'd notes, that thrill the melting heart;
Tuned by thy hand the dulcet harp she rings,
And sounds responsive echo from the strings;
Bright scenes of bliss in trains suggested move,
And charm the world with melody and love.

III.

"Soon the fair forms with vital being bless'd,
Time's feeble children, lose the boon possess'd;
The goaded fibre ceases to obey,   *note*
And sense deserts the uncontractile clay;   340
While births unnumber'd, ere the parents die,

The hourly waste of lovely life supply;
And thus, alternating with death, fulfil
The silent mandates of the Almighty Will;
Whose hand unseen the works of nature dooms
By laws unknown—WHO GIVES, AND WHO RESUMES.

"Each pregnant Oak ten thousand acorns forms
Profusely scatter'd by autumnal storms;
Ten thousand seeds each pregnant poppy sheds *note*
Profusely scatter'd from its waving heads; 350
The countless Aphides, prolific tribe, *note*
With greedy trunks the honey'd sap imbibe; *note*
Swarm on each leaf with eggs or embryons big,
And pendent nations tenant every twig.
Amorous with double sex, the snail and worm,
Scoop'd in the soil, their cradling caverns form;
Heap their white eggs, secure from frost and floods,
And crowd their nurseries with uncounted broods.
Ere yet with wavy tail the tadpole swims, *note*
Breathes with new lungs, or tries his nascent limbs; 360
Her countless shoals the amphibious frog forsakes,
And living islands float upon the lakes.
The migrant herring steers her myriad bands
From seas of ice to visit warmer strands;
Unfathom'd depths and climes unknown explores,
And covers with her spawn unmeasured shores.
—All these, increasing by successive birth,
Would each o'erpeople ocean, air, and earth.

"So human progenies, if unrestrain'd,
By climate friended, and by food sustain'd, 370
O'er seas and soils, prolific hordes! would spread
Erelong, and deluge their terraqueous bed;
But war, and pestilence, disease, and dearth,
Sweep the superfluous myriads from the earth.
Thus while new forms reviving tribes acquire

Each passing moment, as the old expire;
Like insects swarming in the noontide bower,
Rise into being, and exist an hour;
The births and deaths contend with equal strife,
And every pore of Nature teems with Life;   380
Which buds or breathes from Indus to the Poles,   *note*
And Earth's vast surface kindles, as it rolls!

"Hence when a Monarch or a mushroom dies,
Awhile extinct the organic matter lies;
But, as a few short hours or years revolve,
Alchemic powers the changing mass dissolve;
Born to new life unnumber'd insects pant,   *note*
New buds surround the microscopic plant;
Whose embryon senses, and unwearied frames,
Feel finer goads, and blush with purer flames;   390
Renascent joys from irritation spring,
Stretch the long root, or wave the aurelian wing.

"When thus a squadron or an army yields,
And festering carnage loads the waves or fields;
When few from famines or from plagues survive,
Or earthquakes swallow half a realm alive;—
While Nature sinks in Time's destructive storms,
The wrecks of Death are but a change of forms;
Emerging matter from the grave returns,
Feels new desires, with new sensations burns;   400
With youth's first bloom a finer sense acquires,
And Loves and Pleasures fan the rising fires.—
Thus sainted Paul, 'O Death!' exulting cries,   *note*
'Where is thy sting? O Grave! thy victories?'

"Immortal Happiness from realms deceased
Wakes, as from sleep, unlessen'd or increased;
Calls to the wise in accents loud and clear,
Sooths with sweet tones the sympathetic ear;

Informs and fires the revivescent clay,
And lights the dawn of Life's returning day.

"So when Arabia's Bird, by age oppress'd,
Consumes delighted on his spicy nest;
A filial Phoenix from his ashes springs,
Crown'd with a star, on renovated wings;
Ascends exulting from his funeral flame,
And soars and shines, another and the same.

"So erst the Sage with scientific truth
In Grecian temples taught the attentive youth;
With ceaseless change how restless atoms pass
From life to life, a transmigrating mass;
How the same organs, which to day compose
The poisonous henbane, or the fragrant rose,
May with to morrow's sun new forms compile,
Frown in the Hero, in the Beauty smile.
Whence drew the enlighten'd Sage the moral plan,
That man should ever be the friend of man;
Should eye with tenderness all living forms,
His brother-emmets, and his sister-worms.

"Hear, O ye Sons of Time! your final doom,
And read the characters, that mark your tomb:
The marble mountain, and the sparry steep,
Were built by myriad nations of the deep,—
Age after age, who form'd their spiral shells,
Their sea-fan gardens and their coral cells;
Till central fires with unextinguished sway
Raised the primeval islands into day;—
The sand-fill'd strata stretch'd from pole to pole;
Unmeasured beds of clay, and marl, and coal,
Black ore of manganese, the zinky stone,
And dusky steel on his magnetic throne,
In deep morass, or eminence superb,

Rose from the wrecks of animal or herb;
These from their elements by Life combined,
Form'd by digestion, and in glands refined,
Gave by their just excitement of the sense
The Bliss of Being to the vital Ens.

"Thus the tall mountains, that emboss the lands,
Huge isles of rock, and continents of sands,
Whose dim extent eludes the inquiring sight,
Are mighty Monuments of past Delight; *note*
Shout round the globe, how Reproduction strives 451
With vanquish'd Death,—and Happiness survives;
How Life increasing peoples every clime, *note*
And young renascent Nature conquers Time;
—And high in golden characters record
The immense munificence of Nature's Lord!—

"He gives and guides the sun's attractive force,
And steers the planets in their silver course;
With heat and light revives the golden day,
And breathes his spirit on organic clay; 460
With hand unseen directs the general cause
By firm immutable immortal laws."

Charm'd with her words the Muse astonish'd stands,
The Nymphs enraptured clasp their velvet hands;
Applausive thunder from the fane recoils,
And holy echoes peal along the ailes;
O'er Nature's shrine celestial lustres glow,
And lambent glories circle round her brow.

IV.

Now sinks the golden sun,—the vesper song
Demands the tribute of Urania's tongue; 470

Onward she steps, her fair associates calls
From leaf-wove avenues, and vaulted halls.
Fair virgin trains in bright procession move,
Trail their long robes, and whiten all the grove;
Pair after pair to Nature's temple sweep,
Thread the broad arch, ascend the winding steep;
Through brazen gates along susurrant ailes
Stream round their Goddess the successive files;
Curve above curve to golden seats retire,
And star with beauty the refulgent quire. 480

And first to Heaven the consecrated throng
With chant alternate pour the adoring song,
Swell the full hymn, now high, and now profound,
With sweet responsive symphony of sound.
Seen through their wiry harps, below, above,
Nods the fair brow, the twinkling fingers move;
Soft-warbling flutes the ruby lip commands,
And cymbals ring with high uplifted hands.

To Chaos next the notes melodious pass, *note*
How suns exploded from the kindling mass, 490
Waved o'er the vast inane their tresses bright,
And charm'd young Nature's opening eyes with light.
Next from each sun how spheres reluctant burst,
And second planets issued from the first.
And then to Earth descends the moral strain,
How isles, emerging from the shoreless main,
With sparkling streams and fruitful groves began,
And form'd a Paradise for mortal man.

Sublimer notes record Celestial Love,
And high rewards in brighter climes above; 500
How Virtue's beams with mental charm engage
Youth's raptured eye, and warm the frost of age,

Gild with soft lustre Death's tremendous gloom,
And light the dreary chambers of the tomb.
How fell Remorse shall strike with venom'd dart,
Though mail'd in adamant, the guilty heart;
Fierce furies drag to pains and realms unknown
The blood-stain'd tyrant from his tottering throne.

By hands unseen are struck aerial wires,
And Angel-tongues are heard amid the quires; 510
From aile to aile the trembling concord floats,
And the wide roof returns the mingled notes,
Through each fine nerve the keen vibrations dart,
Pierce the charm'd ear, and thrill the echoing heart.—

Mute the sweet voice, and still the quivering strings,
Now Silence hovers on unmoving wings.—
—Slow to the altar fair Urania bends
Her graceful march, the sacred steps ascends,
High in the midst with blazing censer stands,
And scatters incense with illumined hands: 520
Thrice to the Goddess bows with solemn pause,
With trembling awe the mystic veil withdraws,
And, meekly kneeling on the gorgeous shrine,
Lifts her ecstatic eyes to Truth Divine!

End of Canto IV.

NOTES TO CANTO IV.

Blest is the Sage ... (line 7).

> Felix, qui potuit rerum cognoscere causas;
> Quique metus omnes, et inexorabile fatum,
> Subjecit pedibus, strepitumque Acherontis avari.
> [Happy, who had the skill to understand / Nature's hid causes, and beneath his feet / All terrors cast, and death's relentless doom / And the loud roar of greedy Acheron.]
>
> *Virgil: Georgicon.*
> *Book II, line 490.*

The towering eagle ... (line 19).

> Torva leaena lupum sequitur, lupus ipse capellam,
> Florentem cytisum sequitur lasciva capella.
> [The wild lioness pursues the wolf, the wolf chases the goat / the wanton goat seeks the flowering clover.]
>
> *Virgil.*

Fell Oestrus buries ... (line 29). The gadfly, bot-fly, or sheep-fly: the larva lives in the bodies of cattle throughout the whole winter; it is extracted from their backs by an African bird called Buphaga. Adhering to the anus it artfully introduces itself into the intestines of horses, and becomes so numerous in their stomachs, as sometimes to destroy them; it climbs into the nostrils of sheep and calves, and producing a nest of young in a transparent hydatide in the frontal sinus, occasions the vertigo or turn of those animals. In Lapland it so attacks the rein deer that the natives annually travel with the herds from the woods to the mountains. Linneus, Systema Naturae.

The wing'd Ichneumon ... (line 33). Linneus describes seventy-seven species of the ichneumon fly, some of which have a sting as long and some twice as long as their bodies. Many of them insert their eggs into various caterpillars, which when they are hatched seem for a time to prey on the reservoir of silk in the backs of those animals designed for their own use to spin a cord to support them, or a bag to contain them, while they change from their larva form to a butterfly; as I have seen in above fifty cabbage-caterpillars. The ichneumon larva then makes its way out of the caterpillar, and spins itself a small cocoon like a silk worm; these cocoons are about the size of a small pin's head, and I have seen about ten of them on each cabbage caterpillar, which soon dies after their exclusion.

Other species of ichneumon insert their eggs into the aphis, and into the larva of the aphidivorous fly: others into the bedeguar of rose trees, and the gall-nuts of oaks; whence those excrescences seem to be produced, as well as the hydatides in the frontal sinus of sheep and calves by the stimulus of the larvae deposited in them.

While fierce Libellula ... (line 37). The Libellula or Dragon-fly is said to be a most voracious animal; Linneus says in their perfect state they are the hawks to naked winged flies; in their larva state they run beneath the water, and are the cruel crocodiles of aquatic insects. Systema Naturae.

Contending bee-swarms ... (line 39). Stronger bee-swarms frequently attack weak hives, and in two or three days destroy them and carry away their honey; this I once prevented by removing the attacked hive after the first day's battle to a distinct part of the garden. See Phytologia, Sect. XIV, 3, 7.

The shark rapacious ... (line 57). The shark has three rows of sharp teeth within each other, which he can bend downwards internally to admit larger prey, and raise to prevent its return; his snout hangs so far over his mouth, that he is necessitated to turn upon his back, when he takes fish that swim over him, and hence seems peculiarly formed to catch those that swim under him.

The crawling crocodiles ... (line 59). As this animal lives chiefly at the bottom of the rivers, which he frequents, he has the power of opening the upper jaw as well as the under one, and thus with greater facility catches the fish or water-fowl which swim over him.

One great slaughter-house ... (line 66). As vegetables are an inferior order of animals fixed to the soil; and as the locomotive animals prey upon them, or upon each other; the world may indeed be said to be one great slaughter-house. As the digested food of vegetables consists principally of sugar, and from this is produced again their mucilage, starch, and oil, and since animals are sustained by these vegetable productions, it would seem that the sugar-making process carried on in vegetable vessels was the great source of life to all organized beings. And that if our improved chemistry should ever discover the art of making sugar from fossile or aerial matter without the assistance of vegetation, food for animals would then become as plentiful as water, and they might live upon the earth without preying on each other, as thick as blades of grass, with no restraint to their numbers but the want of local room.

It would seem that roots fixed in the earth and leaves innumerable waving in the air were necessary for the decomposition of water and air, and the conversion of them into saccharine matter, which would have been not only cumberous but totally incompatible with the locomotion of animal bodies. For how could a man or quadruped have carried on his head or back a forest of leaves, or have had long branching lacteal or absorbent vessels terminating in the earth? Animals therefore subsist on vegetables; that is they take the matter so prepared, and have organs to prepare it further for the purposes of higher animation and greater sensibility.

While cold and hunger ... (line 71). Those parts of our system, which are in health excited into perpetual action, give us pain, when they are not excited into action: thus when the hands are for a time immersed in snow, an inaction of the cutaneous capillaries is induced, as is seen from the paleness of the skin, which is attended with the pain of coldness. So the pain of hunger is probably produced by the inaction of the muscular fibres of the stomach from the want of the stimulus of food.

Thus those, who have used much voluntary exertion in their early years, and have continued to do so, till the decline of life commences, if they then lay aside their employment, whether that of a minister of state, a general of an army, or a merchant, or manufacturer; they cease to have their faculties excited into their usual activity, and become unhappy, I suppose from the too great accumulation of the sensorial power of volition; which wants the accustomed stimulus or motive to cause its expenditure.

Here laughs Ebriety ... (line 77).

> Saevior armis
> Luxuria incubuit, victumque ulciscitur orbem.
> [Luxury, more deadly than war, broods over the city, and avenges a conquered world.]
>
> Horace. [Eg. Juvenal.]

E'en o'er the grave ... (line 87). Many theatric preachers among the Methodists successfully inculcate the fear of death and of Hell, and live luxuriously on the folly of their hearers: those who suffer under this insanity, are generally most innocent and harmless people, who are then liable to accuse themselves of the greatest imaginary crimes; and have so much intellectual cowardice, that they dare not reason about those things, which they are directed by their priests to believe. Where this intellectual cowardice is great, the voice of reason is ineffectual; but that of ridicule may save many from these mad-making doctors, as the farces of Mr. Foot; though it is too weak to cure those who are already hallucinated.

And last association ... (line 93). The miseries and the felicities of life may be divided into those which arise in consequence of irritation, sensation, volition, and association; and consist in the actions of the extremities of the nerves of sense, which constitute our ideas; if they are much more exerted than usual, or much less exerted than usual, they occasion pain; as when the finger is burnt in a candle; or when we go into a cold bath: while their natural degree of exertion produces the pleasure of life or existence. This pleasure is nevertheless increased, when the system is stimulated into rather stronger action than usual, as after a copious dinner, and at the beginning of intoxication; and diminished, when it is only excited into somewhat less activity than usual, which is termed ennui, or irksomeness of life.

Ideal ills ... (line 94). The tooth-edge is an instance of bodily pain occasioned by association of ideas. Every one in his childhood has repeatedly bit a part of the glass or earthen vessel, in which his food has been given him, and has thence had a disagreeable sensation in his teeth, attended at the same time with a jarring sound: and ever after, when such a sound is accidentally produced, the disagreeable sensation of the teeth follows by association of ideas; this is further elucidated in Zoonomia, Vol. I, Sect. XVI, 10.

Enrich his heir ... (line 100).

Cum furor haud dubius, cum sit manifesta phrenitis,
Ut locuples moriaris, egenti vivere fato.
[It is plain madness and sheer lunacy to live in want because you hope to be wealthy when you die.] *Juvenal.*

A Wolf in wool ... (line 102). A wolf in sheep's clothing.

With airs volcanic ... (line 119). Those epidemic complaints, which are generally termed influenza, are believed to arise from vapours thrown out from earthquakes in such abundance as to affect large regions of the atmosphere, see Botanic Garden, Vol. I, Canto IV, line 65, while the diseases properly termed contagious originate from the putrid effluvia of decomposing animal or vegetable matter.

Sentimental pain ... (line 130). Children should be taught in their early education to feel for all the remediable evils, which they observe in others; but they should at the same time be taught sufficient firmness of mind not intirely to destroy their own happiness by their sympathizing with too great sensibility with the numerous irremediable evils, which exist in the present system of the world: as by indulging that kind of melancholy they decrease the sum total of public happiness; which is so far rather reprehensible than commendable. See Plan for Female Education by Dr. Darwin, Johnson, London, Sect. XVII.

This has been carried to great excess in the East by the disciples of Confucius; the Gentoos during a famine in India refused to eat the flesh of cows and of other animals to satisfy their hunger, and save themselves from death. And at other times they have been said to permit fleas and musquitoes to feed upon them from this erroneous sympathy.

From wandering atoms ... (line 147). Had those ancient philosophers, who contended that the world was formed from atoms, ascribed their combinations to certain immutable properties received from the hand of the Creator, such as general gravitation, chemical affinity, or animal appetency, instead of ascribing them to a blind chance; the doctrine of atoms, as constituting or composing the material world by the variety of their combinations, so far from leading the mind to atheism, would strengthen the demonstration of the existence of a Deity, as the first cause of all things; because the analogy resulting from our perpetual experience of cause and effect would have thus been exemplified through universal nature.

The varied landscape ... (line 160). 1. The pleasure, we feel on examining a fine landscape, is derived from various sources; as first the excitement of the retina of the eye into certain quantities of action; which when there is in the optic nerve any accumulation of sensorial power, is always agreeable. 2. When it is excited into such successive actions, as relieve each other; as when a limb has been long exerted in one direction, by stretching it in another; as described in Zoonomia, Sect. XL, 6, on ocular spectra. 3. And lastly by the associations of its parts with some agreeable sentiments or tastes, as of sublimity, beauty, utility, novelty; and the objects suggesting other sentiments, which have lately been termed picturesque as mentioned in the note to Canto III, line 230 of this work. The two former of these sources of pleasure arise from irritation, the last from association.

We drink delighted ... (line 178). 1. The pleasure we experience from music, is, like that from viewing a landscape, derived from various sources;

as first from the excitement of the auditory nerve into certain quantities of action, when there exists any accumulation of sensorial power. 2. When the auditory nerve is exerted in such successive actions as relieve each other, like stretching or yawning, as described in Botanic Garden, Vol. II, Interlude the third, these successions of sound are termed melody, and their combinations harmony. 3. From the repetition of sounds at certain intervals of time; as we hear them with greater facility and accuracy, when we expect them; because they are then excited by volition, as well as by irritation, or at least the tympanum is then better adapted to assist their production; hence the two musical times or bars; and hence the rhimes in poetry give pleasure, as well as the measure of the verse: and lastly the pleasure we receive from music, arises from the associations of agreeable sentiments with certain proportions, or repetitions, or quantities, or times of sounds which have been previously acquired; as explained in Zoonomia Vol. I, Sect. XVI, 10, and Sect. XXII, 2.

Mark'd the figur'd sand ... (line 242). The ancient orators seem to have spoken disrespectfully of the mechanic philosophers. Cicero mentioning Archimedes, calls him Homunculus e pulvere et radio, alluding to the custom of drawing problems on the sand with a staff.

So Savery guided ... (line 249). Captain Savery first applied the pressure of the atmosphere to raise water in consequence of a vacuum previously produced by the condensation of steam, though the Marquis of Worcester had before proposed to use for this purpose the expansive power of steam; see Botanic Garden, Vol. I, Canto I, line 253, Note.

The waving flax ... (line 254). Flax is said to have been first discovered on the banks of the Nile, and Isis to have been the inventress of spinning and weaving.

So Arkwright taught ... (line 261). See Botanic Garden, Vol. II, Canto II, line 87, Note.

The immortal Press ... (line 270). The discovery of the art of printing has had so great influence on human affairs, that from thence may be dated a new aera in the history of mankind. As by the diffusion of general knowledge, both of the arts of taste and of useful sciences, the public mind has become improved to so great a degree, that though new impositions have been perpetually produced, the arts of detecting them have improved with greater rapidity. Hence since the introduction of printing, superstition has been much lessened by the reformation of religion; and necromancy, astrology, chiromancy, witchcraft, and vampyrism, have vanished from all classes of society; though some are still so weak in the present enlightened times as to believe in the prodigies of animal magnetism, and of metallic tractors; by this general diffusion of knowledge, if the liberty of the press be preserved, mankind will not be liable in this part of the world to sink into such abject slavery as exists at this day in China.

Her expressive verb ... (line 294). The verb, or the word, has been so called from its being the most expressive term in all languages; as it suggests the ideas of existence, action or suffering, and of time; see the Note on Canto III, line 371, of this work.

Call'd by thy voice ... (line 299). The numerous trains of associated ideas are divided by Mr. Hume into three classes, which he has termed contiguity, causation, and resemblance. Nor should we wonder to find them thus connected together, since it is the business of our lives to dispose them into these three classes; and we become valuable to ourselves and our friends as we succeed in it. Those who have combined an extensive class of ideas by the contiguity of time or place, are men learned in the history of mankind, and of the sciences they have cultivated. Those who have connected a great class of ideas of resemblances, possess the source of the ornaments of poetry and oratory, and of all rational analogy. While those who have connected great classes of ideas of causation, are furnished with the powers of producing effects. These are the men of active wisdom who lead armies to victory, and kingdoms to prosperity; or discover and improve the sciences which meliorate and adorn the condition of humanity.

Polish'd wit bestows ... (line 309). Mr. Locke defines wit to consist of an assemblage of ideas, brought together with quickness and variety, wherein can be found any resemblance or congruity, thereby to make up pleasant pictures and agreeable visions in the fancy. To which Mr. Addison adds, that these must occasion surprise as well as delight; Spectator, Vol. I, No. LXII. See Note on Canto III, line 145, and Additional Note VII, 3. Perhaps wit in the extended use of the word may mean to express all kinds of fine writing, as the word Taste is applied to all agreeable visible objects, and thus wit may mean descriptive sublimity, beauty, the pathetic, or ridiculous, but when used in the confined sense, as by Mr. Locke and Mr. Addison as above, it may probably be better defined a combination of ideas with agreeable novelty, as this may be effected by opposition as well as by resemblance.

The goaded fibre ... (line 339). Old age consists in the inaptitude to motion from the inirritability of the system, and the consequent want of fibrous contraction; see Additional Note VII.

Ten thousand seeds ... (line 349). The fertility of plants in respect to seeds is often remarkable; from one root in one summer the seeds of zea, maize, amount to 2000; of inula, elecampane, to 3000; of helianthus, sunflower, to 4000; of papaver, poppy, 32000; of nicotiana, tobacco, to 40320; to this must be added the perennial roots, and the buds. Buds, which are so many herbs, in one tree, the trunk of which does not exceed a span in thickness, frequently amount to 10000; Linneus, Philosophia Botanica, p. 86.

The countless Aphides ... (line 351). The aphises, pucerons, or vine-fretters, are hatched from an egg in the early spring, and are all called females, as they produce a living offspring about once in a fortnight to the ninth generation, which are also all of them females; then males are also produced, and by their intercourse the females become oviparous, and deposite their eggs on the branches, or in the bark to be hatched in the ensuing spring.

This double mode of reproduction, so exactly resembling the buds and seeds of trees, accounts for the wonderful increase of this insect, which, according to Dr. Richardson, consists of ten generations, and of fifty at an average in each generation; so that the sum of fifty multiplied by fifty, and that product again multiplied by fifty nine times, would give the product of one egg only in countless millions; to which must be added the innumerable eggs laid by the tenth generation for the renovation of their progeny in the ensuing spring.

The honey'd sap ... (line 352). The aphis punctures with its fine proboscis the sap-vessels of vegetables without any visible wound, and thus drinks the sap-juice, or vegetable chyle, as it ascends. Hence on the twigs of trees they stand with their heads downwards, as I have observed, to acquire this ascending sap-juice with greater facility. The honey-dew on the upper surface of leaves is evacuated by these insects, as they hang on the underside of the leaves above; when they take too much of this saccharine juice during the vernal or midsummer sap-flow of most vegetables; the black powder on leaves is also their excrement at other times. The vegetable world seems to have escaped total destruction from this insect by the number of flies, which in their larva state prey upon them; and by the ichneumon fly, which deposits its eggs in them. Some vegetables put forth stiff bristles with points round their young shoots, as the moss-rose, apparently to prevent the depredation of these insects, so injurious to them by robbing them of their chyle or nourishment.

The tadpole swims ... (line 359). The progress of a tadpole from a fish to a quadruped by his gradually putting forth his limbs, and at length leaving the water, and breathing the dry air, is a subject of great curiosity, as it resembles so much the incipient state of all other quadrupeds, and men, who are aquatic animals in the uterus, and become aerial ones at their birth.

Which buds or breathes ... (line 381). Organic bodies, besides the carbon, hydrogen, azote, and the oxygen and heat, which are combined with them, require to be also immersed in loose heat and loose oxygen to preserve their mutable existence; and hence life only exists on or near the surface of the earth; see Botanic Garden, Vol. I, Canto IV, line 419.

L'organisation, le sentiment, le movement spontané, la vie, n'existent qu'à la surface de la terre, et dans les lieux exposés à la lumière. Traité de Chimie par M. Lavoisier, Tom. I, p. 202.

Born to new life ... (line 387). From the innumerable births of the larger insects, and the spontaneous productions of the microscopic ones, every part of organic matter from the recrements of dead vegetable or animal bodies, on or near the surface of the earth, becomes again presently reanimated; which by increasing the number and quantity of living organizations, though many of them exist but for a short time, adds to the sum total of terrestrial happiness.

Thus sainted Paul ... (line 403). The doctrine of St. Paul teaches the resurrection of the body in an incorruptible and glorified state, with consciousness of its previous existence; he therefore justly exults over the sting of death, and the victory of the grave.

And lights the dawn ... (line 410). The sum total of the happiness of organized nature is probably increased rather than diminished, when one large old animal dies, and is converted into many thousand young ones; which are produced or supported with their numerous progeny by the same organic matter. Linneus asserts, that three of the flies, called musca vomitoria, will consume the body of a dead horse, as soon as a lion can; Systema Naturae.

So when Arabia's bird ... (line 411). The story of the Phoenix rising from its own ashes with a star upon its head seems to have been an hieroglyphic emblem of the destruction and resuscitation of all things; see Botanic Garden, Vol. I, Canto IV, line 389.

So erst the Sage ... (line 417). It is probable, that the perpetual transmigration of matter from one body to another, of all vegetables and animals, during their lives, as well as after their deaths, was observed by Pythagoras; which he afterwards applied to the soul, or spirit of animation, and taught, that it passed from one animal to another as a punishment for evil deeds, though without consciousness of its previous existence; and from this doctrine he inculcated a system of morality and benevolence, as all creatures thus became related to each other.

The marble mountain ... (line 431). From the increased knowledge in Geology during the present century, owing to the greater attention of philosophers to the situations of the different materials, which compose the strata of the earth, as well as to their chemical properties, it seems clearly to appear, that the nucleus of the globe beneath the ocean consisted of granite; and that on this the great beds of limestone were formed from the shells of marine animals during the innumerable primeval ages of the world; and that whatever strata lie on these beds of limestone, or on the granite, where the limestone does not cover it, were formed after the elevation of islands and continents above the surface of the sea by the recrements of vegetables and of terrestrial animals; see on this subject Botanic Garden, Vol. I, Additional Note XXIV.

Are mighty monuments ... (line 450). The reader is referred to a few pages on this subject in Phytologia, Sect. XIX, 7, 1, where the felicity of organic life is considered more at large; but it is probable that the most certain way to estimate the happiness and misery of organic beings; as it depends on the actions of the organs of sense, which constitute ideas; or of the muscular fibres which perform locomotion; would be to consider those actions, as they are produced or excited by the four sensorial powers of irritation, sensation, volition, and association. A small volume on this subject by some ingenious writer, might not only amuse, as an object of curiosity; but by showing the world the immediate sources of their pains and pleasures might teach the means to avoid the one, and to procure the other, and thus contribute both ways to increase the sum total of organic happiness.

How Life increasing ... (line 453). Not only the vast calcareous provinces, which form so great a part of the terraqueous globe, and also whatever rests upon them, as clay, marl, sand, and coal, were formed from the fluid elements of heat, oxygen, azote, and hydrogen along with carbon, phosphorus, and perhaps a few other substances, which the science of chemistry has not yet decomposed; and gave the pleasure of life to the animals and vegetables, which formed them; and thus constitute monuments of the past happiness of those organized beings. But as those remains of former life are not again totally decomposed, or converted into their original elements, they supply more copious food to the succession of new animal or vegetable beings on their surface; which consists of materials convertible into nutriment with less labour or activity of the digestive powers; and hence the quantity or number of organized bodies, and their improvement in size, as well as their happiness, has been continually increasing, along with the solid parts of the globe; and will probably continue to increase, till the whole terraqueous sphere, and all that inhabit it shall dissolve by a general conflagration, and be again reduced to their elements.

Thus all the suns, and the planets, which circle round them, may again sink into one central chaos; and may again by explosions produce a new world; which in process of time may resemble the present one, and at length again undergo the same catastrophe! these great events may be the result of the immutable laws impressed on matter by the Great Cause of Causes, Parent of Parents, Ens Entium!

To Chaos next ... (line 489).

> Namque canebat uti magnum per inane coacta
> Semina terrarumque, animaeque, marisque fuissent;
> Et liquidi simul ignis; ut his exordia primis
> Omnia, et ipse tener mundi concreverit orbis.
> [For he sang how, through the vast void, the seeds of earth, and air, and sea, and liquid fire withal were gathered together; how from these elements nascent things, yes all, and even the young globe of the world grew together.]
>
> *Virgil: Ecloga VI, line 31.*

ADDITIONAL NOTES

I.
SPONTANEOUS VITALITY OF MICROSCOPIC ANIMALS

> Hence without parent by spontaneous birth
> Rise the first specks of animated earth.
> Canto I, line 247

Prejudices against this doctrine

I. From the misconception of the ignorant or superstitious, it has been thought somewhat profane to speak in favour of spontaneous vital production, as if it contradicted holy writ; which says, that God created animals and vegetables. They do not recollect that God created all things which exist, and that these have been from the beginning in a perpetual state of improvement; which appears from the globe itself, as well as from the animals and vegetables, which possess it. And lastly, that there is more dignity in our idea of the supreme author of all things, when we conceive him to be the cause of causes, than the cause simply of the events, which we see; if there can be any difference in infinity of power!

Another prejudice which has prevailed against the spontaneous production of vitality, seems to have arisen from the misrepresentation of this doctrine, as if the larger animals had been thus produced; as Ovid supposes after the deluge of Deucalion, that lions were seen rising out of the mud of the Nile, and struggling to disentangle their hinder parts.

It was not considered, that animals and vegetables have been perpetually improving by reproduction; and that spontaneous vitality was only to be looked for in the simplest organic beings, as in the smallest microscopic animalcules; which perpetually, perhaps hourly, enlarge themselves by reproduction, like the roots of tulips from seed, or the buds of seedling trees, which die annually, leaving others by solitary reproduction rather more perfect than themselves for many successive years, till at length they acquire sexual organs or flowers.

A third prejudice against the existence of spontaneous vital productions has been the supposed want of analogy; this has also arisen from the expectation, that the larger or more complicated animals should be thus produced; which have acquired their present perfection by successive generations during an uncounted series of ages. Add to this, that the want of analogy opposes the credibility of all new discoveries, as of the magnetic needle, and coated electric jar, and Galvanic pile; which should therefore certainly be well weighed and nicely investigated before distinct credence is given them; but then the want of analogy must at length yield to repeated ocular demonstration.

Preliminary observations

II. Concerning the spontaneous production of the smallest microscopic animals it should be first observed, that the power of reproduction distinguishes organic being, whether vegetable or animal, from inanimate nature. The circulation of fluids in vessels may exist in hydraulic machines, but the power of reproduction belongs alone to life. This reproduction of plants and of animals is of two kinds, which may be termed solitary and sexual. The former of these, as in the reproduction of the buds of trees, and of the bulbs of tulips, and of the polypus, and aphis, appears to be the first or most simple mode of generation, as many of these organic beings afterwards acquire sexual organs, as the flowers of seedling trees, and of seedling tulips, and the autumnal progeny of the aphis. See Phytologia.

Secondly, it should be observed, that by reproduction organic beings are gradually enlarged and improved; which may perhaps more rapidly and uniformly occur in the simplest modes of animated being; but occasionally also in the more complicated and perfect kinds. Thus the buds of a seedling tree, or the bulbs of seedling tulips, become larger and stronger in the second year than the first, and thus improve till they acquire flowers or sexes; and the aphis, I believe, increases in bulk to the eighth or ninth generation, and then produces a sexual progeny. Hence the existence of spontaneous vitality is only to be expected to be found in the simplest modes of animation, as the complex ones have been formed by many successive reproductions.

Experimental facts

III. By the experiments of Buffon, Reaumur, Ellis, Ingenhouz, and others, microscopic animals are produced in three or four days, according to the warmth of the season, in the infusions of all vegetable or animal matter. One or more of these gentlemen put some boiling veal broth into a phial previously heated in the fire, and sealing it up hermetically or with melted wax, observed it to be replete with animalcules in three or four days.

These microscopic animals are believed to possess a power of generating others like themselves by solitary reproduction without sex; and these gradually enlarging and improving for innumerable successive generations. Mr. Ellis in Philosophical Transactions, Vol. LIX, gives drawings of six kinds of animalcula infusoria, which increase by dividing across the middle into two distinct animals. Thus in paste composed of flour and water, which has been suffered to become acescent, the animalcules called eels, Vibrio anguillula, are seen in great abundance; their motions are rapid and strong; they are viviparous, and produce at intervals a numerous progeny: animals similar to these are also found in vinegar; Naturalist's Miscellany by Shaw and Nodder, Vol. II. These eels were probably at first as minute as other microscopic animalcules; but by frequent, perhaps hourly reproduction, have gradually become the large animals above described, possessing wonderful strength and activity.

To suppose the eggs of the former microscopic animals to float in the atmosphere, and pass through the sealed glass phial, is so contrary to apparent nature, as to be totally incredible! and as the latter are viviparous, it is equally absurd to suppose, that their parents float universally in the atmosphere to lay their young in paste or vinegar!

Not only microscopic animals appear to be produced by a spontaneous vital process, and then quickly improve by solitary generation like the buds of trees, or like the polypus and aphis, but there is one vegetable body, which appears to be produced by a spontaneous vital process, and is believed to be propagated and enlarged in so short a time by solitary generation as to become visible to the naked eye; I mean the green matter first attended to by Dr. Priestley, and called by him Conferva fontinalis. The proofs, that this material is a vegetable, are from its giving up so much oxygen, when exposed to the sunshine, as it grows in water, and from its green colour.

Dr. Ingenhouz asserts, that by filling a bottle with well-water, and inverting it immediately into a basin of well-water, this green vegetable is formed in great quantity; and he believes, that the water itself, or some substance contained in the water, is converted into this kind of vegetation, which then quickly propagates itself.

M. Girtanner asserts, that this green vegetable matter is not produced by water and heat alone, but requires the sun's light for this purpose, as he observed by many experiments, and thinks it arises from decomposing water deprived of a part of its oxygen, and laughs at Dr. Priestley for believing that the seeds of this conferva, and the parents of microscopic animals, exist universally in the atmosphere, and penetrate the sides of glass jars; Philosophical Magazine for May 1800.

Besides this green vegetable matter of Dr. Priestley, there is another vegetable, the minute beginnings of the growth of which Mr. Ellis observed by his microscope near the surface of all putrefying vegetable or animal matter, which is the mucor or mouldiness; the vegetation of which was amazingly quick so as to be almost seen, and soon became so large as to be visible to the naked eye. It is difficult to conceive how the seeds of this mucor can float so universally in the atmosphere as to fix itself on all putrid matter in all places.

Theory of Spontaneous Vitality

IV. In animal nutrition the organic matter of the bodies of dead animals, or vegetables, is taken into the stomach, and there suffers decompositions and new combinations by a chemical process. Some parts of it are however absorbed by the lacteals as fast as they are produced by this process of digestion; in which circumstance this process differs from common chemical operations.

In vegetable nutrition the organic matter of dead animals, or vegetables, undergoes chemical decompositions and new combinations on or beneath the surface of the earth; and parts of it, as they are produced, are perpetually absorbed by the roots of the plants in contact with it; in which this also differs from common chemical processes.

Hence the particles which are produced from dead organic matter by chemical decompositions or new consequent combinations, are found proper for the purposes of the nutrition of living vegetable and animal bodies, whether these decompositions and new combinations are performed in the stomach or beneath the soil.

For the purposes of nutrition these digested or decomposed recrements of dead animal or vegetable matter are absorbed by the lacteals of the stomachs of animals or of the roots of vegetables, and carried into the circulation of their blood, and these compose new organic parts to replace others which are destroyed, or to increase the growth of the plant or animal.

It is probable, that as in inanimate or chemical combinations, one of the composing materials must possess a power of attraction, and the other an aptitude to be attracted; so in organic or animated compositions there must be particles with appetencies to unite, and other particles with propensities to be united with them.

Thus in the generation of the buds of trees, it is probable that two kinds of vegetable matter, as they are separated from the solid system, and float in the circulation, become arrested by two kinds of vegetable glands, and are then deposed beneath the cuticle of the tree, and there join together forming a new vegetable, the caudex of which extends from the plumula at the summit to the radicles beneath the soil, and constitutes a single fibre of the bark.

These particles appear to be of two kinds; one of them possessing an appetency to unite with the other, and the latter a propensity to be united with the former; and they are probably separated from the vegetable blood by two kinds of glands, one representing those of the anthers, and the others those of the stigmas, in the sexual organs of vegetables; which is spoken of at large in Phytologia, Sect. VII, and in Zoonomia, Vol. I, Sect. XXXIX, 8, of the third edition, in octavo; where it is likewise shown, that none of these parts which are deposited beneath the cuticle of the tree, is in itself a complete vegetable embryon, but that they form one by their reciprocal conjunction.

So in the sexual reproduction of animals, certain parts separated from the living organs, and floating in the blood, are arrested by the sexual glands of the female, and others by those of the male. Of these none are complete embryon animals, but form an embryon by their reciprocal conjunction.

There hence appears to be an analogy between generation and nutrition, as one is the production of new organization, and the other the restoration of that which previously existed; and which may therefore be supposed to require materials somewhat similar. Now the food taken up by animal lacteals is previously prepared by the chemical process of digestion in the stomach; but that which is taken up by vegetable lacteals, is prepared by chemical dissolution of organic matter beneath the surface of the earth. Thus the particles, which form generated animal embryons, are prepared from dead organic matter by the chemico-animal processes of sanguification and of secretion; while those which form spontaneous microscopic animals or microscopic vegetables are prepared by chemical dissolutions and new combinations of organic matter in watery fluids with sufficient warmth.

It may be here added, that the production and properties of some kinds of inanimate matter, are almost as difficult to comprehend as those of the simplest degrees of animation. Thus the elastic gum, or caoutchouc, and some fossil bitumens, when drawn out to a great length, contract themselves by their elasticity, like an animal fibre by stimulus. The laws of action of these, and all other elastic bodies, are not yet understood; as the laws of the attraction of cohesion, to produce these effects, must be very different from those of general attraction, since the farther the particles of elastic bodies are drawn from each other till they separate, the stronger they seem to attract; and the nearer they are pressed together, the more they seem to repel; as in bending a spring, or in extending a piece of elastic gum; which is the reverse to what occurs in the attractions of disunited bodies; and much wants further investigation. So the spontaneous production of alcohol or of vinegar, by the vinous and acetous fermentations, as well as the production of a mucus by putrefaction which will contract when extended, seems almost as difficult to understand as the spontaneous production of a fibre from decomposing animal or vegetable substances, which will contract when stimulated, and thus constitutes the primordium of life.

Some of the microscopic animals are said to remain dead for many days or weeks, when the fluid in which they existed is dried up, and quickly to recover life and motion by the fresh addition of water and warmth. Thus the Chaos redivivum of Linneus dwells in vinegar and in bookbinders paste: it revives by water after having been dried for years, and is both oviparous and viviparous; Systema Naturae. Thus the vorticella or wheel animal, which is found in rain water that has stood some days in leaden gutters, or in hollows of lead on the tops of houses, or in the slime or sediment left by such water, though it discovers no sign of life except when in the water, yet it is capable of continuing alive for many months though kept in a dry state. In this state it is of a globulous shape, exceeds not the bigness of a grain of sand, and no signs of life appear; but being put into water, in the space of half an hour a languid motion begins, the globule turns itself about, lengthens itself by slow degrees, assumes the form of a lively maggot, and most commonly in a few minutes afterwards puts out its wheels, swimming vigorously through the water as if in search of food; or else, fixing itself by the tail, works the wheels in such a manner as to bring its food to its mouth; English Encyclopedia, Art. Animalcule.

Thus some shell-snails in the cabinets of the curious have been kept in a dry state for ten years or longer, and have revived on being moistened with warmish water; Philosophical Transactions. So eggs and seeds after many months torpor, are revived by warmth and moisture; hence it may be concluded, that even the organic particles of dead animals may, when exposed to a due degree of warmth and moisture, regain some degree of vitality, since this is done by more complicate animal organs in the instances above mentioned.

The hydra of Linneus, which dwells in the rivers of Europe under aquatic plants, has been observed by the curious of the present time, to revive after it has been dried, to be restored after being mutilated, to multiply by being divided, to be propagated from small portions, to live after being inverted; all which would be best explained by the doctrine of spontaneous reproduction from organic particles not yet completely decomposed.

To this should be added, that these microscopic animals are found in all solutions of vegetable or animal matter in water; as black pepper steeped in water, hay suffered to become putrid in water, and the water of dunghills, afford animalcules in astonishing numbers. See Mr. Ellis's curious account of Animalcules produced from an infusion of Potatoes and Hempseed; Philosophical Transactions, Vol. LIX, from all which it would appear, that organic particles of dead vegetables and animals during their usual chemical changes into putridity or acidity, do not lose all their organization or vitality, but retain so much of it as to unite with the parts of living animals in the process of nutrition, or unite and produce new complicate animals by secretion as in generation, or produce very simple microscopic animals or microscopic vegetables, by their new combinations in warmth and moisture.

And finally, that these microscopic organic bodies are multiplied and enlarged by solitary reproduction without sexual intercourse till they acquire greater perfection or new properties. Lewenhoek observed in rain-water which had stood a few days, the smallest scarcely visible microscopic animalcules, and in a few more days he observed others eight times as large; English Encyclopedia, Art. Animalcule.

Conclusion

There is therefore no absurdity in believing that the most simple animals and vegetables may be produced by the congress of the parts of decomposing organic matter, without what can properly be termed generation, as the genus did not previously exist; which accounts for the endless varieties, as well as for the immense numbers of microscopic animals.

The green vegetable matter of Dr. Priestley, which is universally produced in stagnant water, and the mucor, or mouldiness, which is seen on the surface of all putrid vegetable and animal matter, have probably no parents, but a spontaneous origin from the congress of the decomposing organic particles, and afterwards propagate themselves. Some other fungi, as those growing in close wine-vaults, or others which arise from decaying trees, or rotten timber, may perhaps be owing to a similar spontaneous production, and not previously exist as perfect organic beings in the juices of the wood, as some have supposed. In the same manner it would seem, that the common esculent mushroom is produced from horse dung at any time and in any place, as is the common practice of many gardeners; Kennedy on Gardening.

Appendix

The knowledge of microscopic animals is still in its infancy: those already known are arranged by Mr. Muller into the following classes; but it is probable, that many more classes, as well as innumerable individuals, may be discovered by improvements of the microscope, as Mr. Herschell has discovered so many thousand stars, which were before invisible, by improvements of the telescope.

Mr. Muller's classes consist of

I. *Such as have no External Organs.*

1. Monas: Punctiformis. A mere point.
2. Proteus: Mutabilis. Mutable.
3. Volvox: Sphaericum. Spherical.
4. Enchelis: Cylindracea. Cylindrical.
5. Vibrio: Elongatum. Long.

- Membranaceous.

6. Cyclidium: Ovale. Oval.
7. Paramecium: Oblongum. Oblong.
8. Kolpoda: Sinuatum. Sinuous.
9. Gonium: Angulatum. With angles.
10. Bursaria. Hollow like a purse.

II. *Those that have External Organs.*

- Naked, or not enclosed in a shell:
1. Cercaria: Caudatum. With a tail.
2. Trichoda: Crinitum. Hairy.
3. Kerona: Corniculatum. With horns.
4. Himantopus: Cirratum. Cirrated.
5. Leucophra: Ciliatum undique. Every part ciliated.
6. Vorticella: Ciliatum apice. The apex ciliated.

- Covered with a shell:
7. Brachionus: Ciliatum apice. The apex ciliated.

1. These animalcules are discovered in two or three days in all decompositions of organic matter, whether vegetable or animal, in moderate degrees of warmth with sufficient moisture.

2. They appear to enlarge in a few days, and some to change their form; which are probably converted from more simple into more complicate animalcules by repeated reproductions. See Note VIII.

3. In their early state they seem to multiply by viviparous solitary reproduction, either by external division, as the smaller ones, or by an internal progeny, as the eels in paste or vinegar; and lastly, in their more mature state, the larger ones are said to appear to have sexual connexion. English Encyclopedia.

4. Those animalcules discovered in pustules of the itch, in the feces of dysenteric patients, and in semine masculino, I suppose to be produced by the stagnation and incipient decomposition of those materials in their receptacles, and not to exist in the living blood or recent secretions; as none, I believe, have been discovered in blood when first drawn from the arm, or in fluids newly secreted from the glands, which have not previously stagnated in their reservoirs.

5. They are observed to move in all directions with ease and rapidity, and to avoid obstacles, and not to interfere with each other in their motions. When the water is in part evaporated, they are seen to flock towards the remaining part, and show great agitation. They sustain a great degree of cold, as some insects, and perish in much the same degree of heat as destroys insects; all which evince that they are living animals.

And it is probable, that other or similar animalcules may be produced in the air, or near the surface of the earth, but it is not so easy to view them as in water; which as it is transparent, the creatures produced in it can easily be observed by applying a drop to a microscope. I hope that microscopic researches may again excite the attention of philosophers, as unforeseen advantages may probably be derived from them, like the discovery of a new world.

II.
THE FACULTIES OF THE SENSORIUM

Next the long nerves unite their silver train,
And young Sensation permeates the brain.
Canto I, line 269

I. The fibres, which constitute the muscles and organs of sense, possess a power of contraction. The circumstances attending the exertion of this power of contraction constitute the laws of animal motion, as the circumstances attending the exertion of the power of attraction constitute the laws of motion of inanimate matter.

II. The spirit of animation is the immediate cause of the contraction of animal fibres, it resides in the brain and nerves, and is liable to general or partial diminution or accumulation.

III. The stimulus of bodies external to the moving organ is the remote cause of the original contractions of animal fibres.

IV. A certain quantity of stimulus produces irritation, which is an exertion of the spirit of animation exciting the fibres into contraction.

V. A certain quantity of contraction of animal fibres, if it be perceived at all, produces pleasure; a greater or less quantity of contraction, if it be perceived at all, produces pain; these constitute sensation.

VI. A certain quantity of sensation produces desire or aversion; these constitute volition.

VII. All animal motions which have occurred at the same time, or in immediate succession, become so connected, that when one of them is reproduced, the other has a tendency to accompany or succeed it. When fibrous contractions succeed or accompany other fibrous contractions, the connexion is termed association; when fibrous contractions succeed sensorial motions, the connexion is termed causation; when fibrous and sensorial motions reciprocally introduce each other, it is termed catenation of animal motions.

VIII. These four faculties of the sensorium during their inactive state are termed irritability, sensibility, voluntarily, and associability; in their active state they are termed as above irritation, sensation, volition, association.

Irritation is an exertion or change of some extreme part of the sensorium residing in the muscles or organs of sense, in consequence of the appulses of external bodies.

Sensation is an exertion or change of the central parts of the sensorium, or of the whole of it, beginning at some of those extreme parts of it, which reside in the muscles or organs of sense.

Volition is an exertion or change of the central parts of the sensorium, or of the whole of it, terminating in some of those extreme parts of it, which reside in the muscles or organs of sense.

Association is an exertion or change of some extreme part of the sensorium residing in the muscles or organs of sense, in consequence of some antecedent or attendant fibrous contractions; see Zoonomia, Vol. I.

The word sensorium is used to express not only the medullary part of the brain, spinal marrow, nerves, organs of sense and muscles, but also at the same time that living principle, or spirit of animation, which resides throughout the body, without being cognizable to our senses except by its effects.

III.
VOLKANOES

> Next when imprison'd fires in central caves
> Burst the firm earth, and drank the headlong waves.
> Canto I, line 321

The great and repeated explosions of volcanoes are shown by Mr. Mitchell in the Philosophical Transactions to arise from their communication with the sea, or with rivers, or inundations; and that after a chink or crack is made, the water rushing into an immense burning cavern, and falling on boiling lava, is instantly expanded into steam, and produces irresistible explosions.

As the first volcanic fires had no previous vent, and were probably more central, and larger in quantity, before they burst the crust of the earth then intire, and as the sea covered the whole, it must rapidly sink down into every opening chink; whence these primeval earthquakes were of much greater extent, and of much greater force, than those which occur in the present era.

It should be added, that there may be other elastic vapours produced by great heat from whatever will evaporate, as mercury, and even diamonds; which may be more elastic, and consequently exert greater force than the steam of water even though heated red hot. Which may thence exert a sufficient power to raise islands and continents, and even to throw the moon from the earth.

If the moon be supposed to have been thus thrown out of the great cavity which now contains the South Sea, the immense quantity of water flowing in from the primeval ocean, which then covered the earth, would much contribute to leave the continents and islands, which might be raised at the same time above the surface of the water. In later days there are accounts of large stones falling from the sky, which may have been thus thrown by explosion from some distant earthquake, without sufficient force to cause them to circulate round the earth, and thus produce numerous small moons or satellites.

Mr. Mitchell observes, that the agitations of the earth from the great earthquake at Lisbon were felt in this country about the same time after the shock, as sound would have taken in passing from Lisbon hither; and thence ascribes these agitations to the vibrations of the solid earth, and not to subterraneous caverns of communication; Philosophical Transactions. But from the existence of warm springs at Bath and Buxton, there must certainly be unceasing subterraneous fires at some great depth beneath those parts of this island; see on this subject Botanic Garden, Vol. II, Canto IV, line 79, Note. For an account of the noxious vapours emitted from volcanoes, see Botanic Garden, Vol. II, Canto IV, line 328, Note. For the milder effects of central fires, see Botanic Garden, Vol. I, Canto I, line 139, and Additional Note VI.

IV.

MUSQUITO

So from deep lakes the dread musquito springs,
Drinks the soft breeze, and dries his tender wings.
Canto I, line 347

The gnat, or musquito, Culex pipiens. The larva of this insect lives chiefly in water, and the pupa moves with great agility. It is fished for by ducks; and, when it becomes a fly, is the food of the young of partridges, quails, sparrows, swallows, and other small birds. The females wound us, and leave a red point; and in India their bite is more venomous. The male has its antennae and feelers feathered, and seldom bites or sucks blood; Linneus, Systema Naturae.

It may be driven away by smoke, especially by that from inula helenium, elecampane; and by that of cannabis, hemp. Kalm. It is said that a light in a chamber will prevent their attack on sleeping persons.

The gnats of this country are produced in greater numbers in some years than others, and are then seen in swarms for many evenings near the lakes or rivers whence they arise; and, I suppose, emigrate to upland situations, where fewer of them are produced. About thirty years ago such a swarm was observed by Mr. Whitehurst for a day or two about the lofty tower of Derby church, as to give a suspicion of the fabric being on fire.

Many other kinds of flies have their origin in the water, as perhaps the whole class of neuroptera. Thus the libellula, dragon fly: the larva of which hurries amid the water, and is the cruel crocodile of aquatic insects. After they become flies, they prey principally on the class of insects termed lepidoptera, and diptera of Linneus. The ephemera is another of this order, which rises from the lakes in such quantities in some countries, that the rustics have carried cart-loads of them to manure their corn lands; the larva swims in the water: in its fly-state the pleasures of life are of short duration, as its marriage, production of its progeny, and funeral, are often celebrated in one day. The phryganea is another fly of this order; the larva lies concealed under the water in moveable cylindrical tubes of their own making. In the fly-state they institute evening dances in the air in swarms, and are fished for by the swallows.

Many other flies, who do not leave their eggs in water, contrive to lay them in moist places, as the oestros bovis; the larvae of which exist in the bodies of cattle, where they are nourished during the winter, and are occasionally extracted by a bird of the crow-kind called buphaga. These larvae are also found in the stomachs of horses, whom they sometimes destroy; another species of them adhere to the anus of horses, and creep into the lowest bowel, and are called botts; and another species enters the frontal sinus of sheep, occasioning a vertigo called the turn.

The Musca pendula lives in stagnant water; the larva is suspended by a thread-form respiratory tube; of the musca chamaeleon, the larva lives in fountains, and the fly occasionally walks upon the water. The Musca vomitoria is produced in carcases; three of these flies consume the dead body of a horse as soon as a lion. Linneus, Systema Naturae.

V.
AMPHIBIOUS ANIMALS

> So still the Diodons, amphibious tribe,
> With twofold lungs the sea and air imbibe.
> Canto I, line 351

D.D. Garden dissected the amphibious creature called diodon by Linneus, and was amazed to find that it possessed both external gills and internal lungs, which he described and prepared and sent to Linneus; who thence put this animal into the order nantes of his class amphibia. He adds also, in his account of polymorpha before the class amphibia, that some of this class breathe by lungs only, and others by both lungs and gills.

Some amphibious quadrupeds, as the beaver, water rat, and otter, are said to have the foramen ovale of the heart open, which communicates from one cavity of it to the other; and that, during their continuance under water, the blood can thus for a time circulate without passing through the lungs; but as it cannot by these means acquire oxygen either from the air or water, these creatures find it frequently necessary to rise to the surface to respire. As this foramen ovale is always open in the foetus of quadrupeds, till after its birth that it begins to respire, it has been proposed by some to keep young puppies three or four times a day for a minute or two under warm water to prevent this communication from one cavity of the heart to the other from growing up; whence it has been thought such dogs might become amphibious. It is also believed that this circumstance has existed in some divers for pearl; whose children are said to have been thus kept under water in their early infancy to enable them afterwards to succeed in their employment.

But the most frequent distinction of the amphibious animals, that live much in the water, is, that their heart consists but of one cell; and as they are pale creatures with but little blood, and that colder and darker coloured, as frogs and lizards, they require less oxygen than the warmer animals with a greater quantity and more scarlet blood; and thence, though they have only lungs, they can stay long under water without great inconvenience; but are all of them, like frogs, and crocodiles, and whales, necessitated frequently to rise above the surface for air.

In this circumstance of their possessing a one-celled heart, and colder and darker blood, they approach to the state of fish; which thus appear not to acquire so much oxygen by their gills from the water as terrestrial animals do by their lungs from the atmosphere; whence it may be concluded that the gills of fish do not decompose the water which passes through them, and which contains so much more oxygen than the air, but that they only procure a small quantity of oxygen from the air which is diffused in the water; which also is further confirmed by an experiment with the air-pump, as fish soon die when put in a glass of water into the exhausted receiver, which they would not do if their gills had power to decompose the water and obtain the oxygen from it.

The lamprey, petromyzon, is put by Linneus amongst the nantes, which are defined to possess both gills and lungs. It has seven spiracula, or breathing holes, on each side of the neck, and by its more perfect lungs approaches to the serpent kind; Systema Naturae. The means by which it adheres to stones, even in rapid streams, is probably owing to a partial vacuum made by its respiring organs like sucking, and may be compared to the ingenious method by which boys are seen to lift large stones in the street, by applying to them a piece of strong moist leather with a string through the centre of it; which, when it is forcibly drawn upwards, produces a partial vacuum under it, and thus the stone is supported by the pressure of the atmosphere.

The leech, hirudo, and the remora, echeneis, adhere strongly to objects probably by a similar method. I once saw ten or twelve leeches adhere to each foot of an old horse a little above his hoofs, who was grazing in a morass, and which did not lose their hold when he moved about. The bare-legged travellers in Ceylon are said to be much infested by leeches; and the sea-leech, hirudo muricata, is said to adhere to fish, and the remora is said to adhere to ships in such numbers as to retard their progress.

The respiratory organ of the whale, I suppose, is pulmonary in part, as he is obliged to come frequently to the surface, whence he can be pursued after he is struck with the harpoon; and may nevertheless be in part like the gills of other fish, as he seems to draw in water when he is below the surface, and emits it again when he rises above it.

VI.
HIEROGLYPHIC CHARACTERS

So erst, as Egypt's rude designs explain.
Canto I, line 371

The outlines of animal bodies, which gave names to the constellations, as well as the characters used in chemistry for the metals, and in astronomy for the planets, were originally hieroglyphic figures, used by the magi of Egypt before the invention of letters, to record their discoveries in those sciences.

Other hieroglyphic figures seem to have been designed to perpetuate the events of history, the discoveries in other arts, and the opinions of those ancient philosophers on other subjects. Thus their figures of Venus for beauty, Minerva for wisdom, Mars and Bellona for war, Hercules for strength, and many others, became afterwards the deities of Greece and Rome; and together with the figures of Time, Death, and Fame, constitute the language of the painters to this day.

From the similarity of the characters which designate the metals in chemistry, and the planets in astronomy, it may be concluded that these parts of science were then believed to be connected; whence astrology seems to have been a very early superstition. These, so far, constitute an universal visible language in those sciences.

So the glory, or halo, round the head is a part of the universal language of the eye, designating a holy person; wings on the shoulders denote a good angel; and a tail and hoof denote the figure of an evil demon; to which may be added the cap of liberty and the tiara of popedom. It is to be wished that many other universal characters could be introduced into practice, which might either constitute a more comprehensive language for painters, or for other arts; as those of ciphers and signs have done for arithmetic and algebra, and crotchets for music, and the alphabets for articulate sounds; so a zigzag line made on white paper by a black-lead pencil, which communicates with the surface of the mercury in the barometer, as the paper itself is made constantly to move laterally by a clock, and daily to descend through the space necessary, has ingeniously produced a most accurate visible account of the rise and fall of the mercury in the barometer every hour in the year.

Mr. Grey's Memoria Technica was designed as an artificial language to remember numbers, as of the eras, or dates of history. This was done by substituting one consonant and one vowel for each figure of the ten cyphers used in arithmetic, and by composing words of these letters; which words Mr. Grey makes into hexameter verses, and produces an audible jargon, which is to be committed to memory, and occasionally analysed into numbers when required. An ingenious French botanist, Monsieur Bergeret, has proposed to apply this idea of Mr. Grey to a botanical nomenclature, by making the name of each plant to consist of letters, which, when analysed, were to signify the number of the class, order, genus, and species, with a description also of some particular part of the plant, which was designed to be both an audible and visible language.

Bishop Wilkins in his elaborate "Essay towards a Real Character and a Philosophical Language," has endeavoured to produce, with the greatest simplicity, and accuracy, and conciseness, an universal language both to be written and spoken, for the purpose of the communication of all our ideas with greater exactness and less labour than is done in common languages, as they are now spoken and written. But we have to lament that the progress of general science is yet too limited both for his purpose, and for that even of a nomenclature for botany; and that the science of grammar, and even the number and manner of the pronunciation of the letters of the alphabet, are not yet determined with such accuracy as would be necessary to constitute Bishop Wilkins's grand design of an universal language, which might facilitate the acquirement of knowledge, and thus add to the power and happiness of mankind.

VII.
OLD AGE AND DEATH

> The age-worn fibres goaded to contract
> By repetition palsied, cease to act.
>
> Canto II, line 3

I. *Effects of Age*

The immediate cause of the infirmities of age, or of the progress of life to death, has not yet been well ascertained. The answer to the question, why animals become feeble and diseased after a time, though nourished with the same food which increased their growth from infancy, and afterwards supported them for many years in unimpaired health and strength, must be sought for from the laws of animal excitability, which, though at first increased, is afterwards diminished by frequent repetitions of its adapted stimulus, and at length ceases to obey it.

1. There are four kinds of stimulus which induce the fibres to con-

tract, which constitute the muscles or the organs of sense; as, first, The application of external bodies, which excites into action the sensorial power of irritation; 2dly, Pleasure and pain, which excite into action the sensorial power of sensation; 3dly, Desire and aversion, which excite into action the power of volition; and lastly, The fibrous contractions, which precede association, which is another sensorial power; see Zoonomia, Vol. I, Sect. II, 13.

Many of the motions of the organic system, which are necessary to life, are excited by more than one of these stimuli at the same time, and some of them occasionally by them all. Thus respiration is generally caused by the stimulus of blood in the lungs, or by the sensation of the want of oxygen; but is also occasionally voluntary. The actions of the heart also, though generally owing to the stimulus of the blood, are also inflamed by the association of its motions with those of the stomach, whence sometimes arises an inequality of the pulse, and with other parts of the system, as with the capillaries, whence heat of the skin in fevers with a feeble pulse, see Zoonomia. They are also occasionally influenced by sensation, as is seen in the paleness occasioned by fear, or the blush of shame and anger; and lastly the motions of the heart are sometimes assisted by volition; thus in those who are much weakened by fevers, the pulse is liable to stop during their sleep, and to induce great distress; which is owing at that time to the total suspension of voluntary power; the same occurs during sleep in some asthmatic patients.

2. The debility of approaching age appears to be induced by the inactivity of many parts of the system, or their disobedience to their usual kinds and quantities of stimulus: thus the pallid appearance of the skin of old age is owing to the inactivity of the heart, which ceases to obey the irritation caused by the stimulus of the blood, or its association with other moving organs with its former energy; whence the capillary arteries are not sufficiently distended in their diastole, and consequently contract by their elasticity, so as to close the canal, and their sides gradually coalesce. Of these, those which are most distant from the heart, and of the smallest diameters, will soonest close, and become impervious; hence the hard pulse of aged patients is occasioned by the coalescence of the sides of the vasa vasorum, or capillary arteries of the coats of the other arteries.

The veins of elderly people become turgid or distended with blood, and stand prominent on the skin; for as these do not possess the elasticity of the arteries, they become distended with accumulation of blood; when the heart by its lessened excitability does not contract sufficiently forcibly, or frequently, to receive, as fast as usual, the returning blood; and their apparent prominence on the skin is occasioned by the deficient secretion of fat or mucus in the cellular membrane; and also to the contraction and coalescence and consequent less bulk of many capillary arteries.

3. Not only the muscular fibres lose their degree of excitability from age, as in the above examples; and as may be observed in the tremulous hands and feeble step of elderly persons; but the organs of sense become less excitable by the stimulus of external objects; whence the sight and hearing become defective; the stimulus of the sensorial power of sensation also less affects the aged, who grieve less for the loss of friends or for other disappointments; it should nevertheless be observed, that when the sensorial power of irritation is much exhausted, or its production much diminished; the sensorial power of sensation appears for a time to be increased; as in intoxication there exists a kind of delirium and quick flow of ideas, and yet the person becomes so weak as to totter as he walks; but this delirium is owing to the defect of voluntary power to correct the streams of ideas by intuitive analogy, as in dreams: see Zoonomia: and thus also those who are enfeebled by habits of much vinous potation, or even by age alone, are liable to weep at shaking hands with a friend, whom they have not lately seen; which is owing to defect of voluntary power to correct their trains of ideas caused by sensation, and not to the increased quantity of sensation, as I formerly supposed.

The same want of voluntary power to keep the trains of sensitive ideas consistent, and to compare them by intuitive analogy with the order of nature, is the occasion of the starting at the clapping to of a door, or the fall of a key, which occasions violent surprise with fear and sometimes convulsions, in very feeble hysterical patients, and is not owing I believe (as I formerly supposed) to increased sensation; as they are less sensible to small stimuli than when in health.

Old people are less able also to perform the voluntary exertions of exercise or of reasoning, and lastly the association of their ideas becomes more imperfect, as they are forgetful of the names of persons and places; the associations of which are less permanent, than those of the other words of a language, which are more frequently repeated.

4. This disobedience of the fibres of age to their usual stimuli, has generally been ascribed to repetition or habit, as those who live near a large clock, or a mill, or a waterfall, soon cease to attend to the perpetual noise of it in the day, and sleep dining the night undisturbed. Thus all medicines, if repeated too frequently, gradually lose their effect; as wine and opium cease to intoxicate: some disagreeable tastes as tobacco, by frequent repetition cease to be disagreeable; grief and pain gradually diminish and at length cease altogether; and hence life itself becomes tolerable.

This diminished power of contraction of the fibres of the muscles or organs of sense, which constitutes permanent debility or old age, may arise from a deficient secretion of sensorial power in the brain, as well as from the disobedience of the muscles and organs of sense to their usual stimuli; but this less production of sensorial power must depend on the inactivity of the glands, which compose the brain, and are believed to separate it perpetually from the blood; and is thence owing to a similar cause with the inaction of the fibres of the other parts of the system.

It is finally easy to understand how the fibres may cease to act by the usual quantity of stimulus after having been previously exposed to a greater quantity of stimulus, or to one too long continued; because the expenditure of sensorial power has then been greater than its production; but it is not easy to explain why the repetition of fibrous contractions, which during the meridian of life did not expend the sensorial power faster than it was produced; or only in such a degree as was daily restored by rest and sleep, should at length in the advance of life expend too much of it; or otherwise, that less of it should be produced in the brain; or reside in the nerves; lastly that the fibres should become less excitable by the usual quantity of it.

5. But these facts would seem to show, that all parts of the system are not changed as we advance in life, as some have supposed; as in that case it might have preserved for ever its excitability; and it might then perhaps have been easier for nature to have continued her animals and vegetables for ever in their mature state, than perpetually by a complicate

apparatus to have produced new ones, and suffer the old ones to perish; for a further account of stimulus and the consequent animal exertion, see Zoonomia, Vol. I, Sect. 12.

II. *Means of preventing old age*

The means of preventing the approach of age must therefore consist in preventing the inexcitability of the fibres, or the diminution of the production of sensorial power.

1. As animal motion cannot be performed without the fluid matter of heat, in which all things are immersed, and without a sufficient quantity of moisture to prevent rigidity: nothing seems so well adapted to both these purposes as the use of the warm bath; and especially in those, who become thin or emaciated with age, and who have a hard and dry skin, with hardness of the coat of the arteries; which feels under the finger like a cord; the patient should sit in warm water for half an hour every day, or alternate days, or twice a week; the heat should be about ninety-eight degrees on Fahrenheit's scale, or of such a warmth, as may be most agreeable to his sensation; but on leaving the bath he should always be kept so cool, whether he goes into bed, or continues up, as not sensibly to perspire.

There is a popular prejudice, that the warm bath relaxes people, and that the cold bath braces them; which are mechanical terms belonging to drums and fiddle-strings, but not applicable except metaphorically to animal bodies, and then commonly mean weakness and strength: during the continuance in the bath the patient does not lose weight, unless he goes in after a full meal, but generally weighs heavier as the absorption is greater than the perspiration; but if he suffers himself to sweat on his leaving the bath, he will undoubtedly be weakened by the increased action of the system, and its exhaustion: the same occurs to those who are heated by exercise, or by wine, or spice, but not during their continuance in the warm bath: whence we may conclude, that the warm bath is the most harmless of all those stimuli, which are greater than our natural habits have accustomed us to; and that it particularly counteracts the approach of old age in emaciated people with dry skins.

It may be here observed in favour of bathing, that some fish are believed to continue to a great age, and continually to enlarge in size, as they advance in life; and that long after their state of puberty. I have seen perch full of spawn, which were less than two inches long; and it is known, that they will grow to six or eight times that size; it is said, that the whales, which have been caught of late years, are much less in size than those, which were caught, when first the whale-fishery was established; as the large ones, which were supposed to have been some hundred years old, are believed to be already destroyed.

All cold-blooded amphibious animals more slowly waste their sensorial power; as they are accustomed to less stimulus from their respiring less oxygen; and their movements in water are slower than those of aerial animals from the greater resistance of the element. There besides seems to be no obstacle to the growth of aquatic animals; as by means of the air-bladder, they can make their specific gravity the same as that of the water in which they swim. And the moisture of the element seems well adapted to counteract the rigidity of their fibres; and as their exertions in locomotion, and the pressure of some parts on others, are so much less than in the bodies of land animals.

2. But as all excessive stimuli exhaust the sensorial power, and render the system less excitable for a time till the quantity of sensorial power is restored by sleep, or by the diminution or absence of stimulus; which is seen by the weakness of inebriates for a day at least after intoxication. And as the frequent repetition of this great and unnatural stimulus of fermented liquors produces a permanent debility, or disobedience of the system to the usual and natural kinds and quantities of stimulus, as occurs in those who have long been addicted to the ingurgitation of fermented liquors.

And as, secondly, the too great deficiency of the quantity of natural stimuli, as of food, and warmth, or of fresh air, produces also diseases; as is often seen in the children of the poor in large towns, who become scrofulous from want of due nourishment, and from cold, damp, unairy lodgings.

The great and principal means to prevent the approach of old age and death, must consist in the due management of the quantity of every kind of stimulus, but particularly of that from objects external to the moving organ; which may excite into action too great or too small a quantity of the sensorial power of irritation, which principally actuates the vital organs. Whence the use of much wine, or opium, or spice, or of much salt, by their unnatural stimulus induces consequent debility, and shortens life, on the one hand, by the exhaustion of sensorial power; so on the other hand, the want of heat, food, and fresh air, induces debility from defect of stimulus, and a consequent accumulation of sensorial power, and a general debility of the system. Whence arise the pains of cold and hunger, and those which are called nervous; and which are the cause of hysteric, epileptic, and perhaps of asthmatic paroxysms, and of the cold fits of fever.

3. Though all excesses of increase and decrease of stimulus should be avoided, yet a certain variation of stimulus seems to prolong the excitability of the system; as during any diminution of the usual quantity of stimulus, an accumulation of sensorial power is produced; and in consequence the excitability, which was lessened by the action of habitual stimulus, becomes restored. Thus those, who are uniformly habituated to much artificial heat, as in warm parlours in the winter months, lose their irritability in some degree, and become feeble like hot-house plants; but by frequently going for a time into the cold air, the sensorial power of irritability is accumulated and they become stronger.

Whence it may be deduced, that the variations of the cold and heat of this climate contribute to strengthen its inhabitants, who are more active and vigorous, and live longer, than those of either much warmer or much colder latitudes.

This accumulation of sensorial power from diminution of stimulus any one may observe, who in severe weather may sit by the fire-side till he is chill and uneasy with the sensation of cold; but if he walks into the frosty air for a few minutes, an accumulation of sensorial power is produced by diminution of the stimulus of heat, and on his returning into the room where he was chill before, his whole skin will now glow with warmth.

Hence it may be concluded, that the variations of the quantity of stimuli within certain limits contribute to our health; and that those houses which are kept too uniformly warm, are less wholesome than where the inhabitants are occasionally exposed to cold air in passing from one room to another.

Nevertheless to those weak habits with pale skins and large pupils of the eyes, whose degree of irritability is less than health requires, as in scrofulous, hysterical, and some consumptive constitutions, a climate warmer than our own may be of service, as a greater stimulus of heat may be wanted to excite their less irritability. And also a more uniform quantity of heat may be serviceable to consumptive patients than is met with in this country, as the lungs cannot be clothed like the external skin, and are therefore subject to greater extremes of heat and cold in passing in winter from a warm room into the frosty air.

4. It should nevertheless be observed, that there is one kind of stimulus, which though it be employed in quantity beyond its usual state, seems to increase the production of sensorial power beyond the expenditure of it (unless its excess is great indeed) and thence to give permanent strength and energy to the system; I mean that of volition. This appears not only from the temporary strength of angry or insane people, but because insanity even cures some diseases of debility, as I have seen in dropsy, and in some fevers; but it is also observable, that many who have exerted much voluntary effort during their whole lives, have continued active to great age. This however may be conceived to arise from these great exertions being performed principally by the organs of sense, that is by exciting and comparing ideas; as in those who have invented sciences, or have governed nations, and which did not therefore exhaust the sensorial power of those organs which are necessary to life, but perhaps rather prevented them from being sooner impaired, their sensorial power not having been so frequently exhausted by great activity, for very violent exercise of the body, long continued, forwards old age; as is seen in post-horses that are cruelly treated, and in many of the poor, who with difficulty support their families by incessant labour.

III. *Theory of the Approach of Age*

The critical reader is perhaps by this time become so far interested in this subject as to excuse a more prolix elucidation of it.

In early life the repetition of animal actions occasions them to be performed with greater facility, whether those repetitions are produced by volition, sensation, or irritation; because they soon become associated together, if as much sensorial power is produced between every reiteration of action, as is expended by it.

But if a stimulus be repeated at uniform intervals of time, the action, whether of our muscles or organs of sense, is performed with still greater facility and energy; because the sensorial power of association mentioned above, is combined with the sensorial power of irritation, and forms part of the diurnal chain of animal motions; that is, in common language, the acquired habit assists the power of the stimulus; see Zoonomia, Vol. I, Sect. XXII, 2, and Sect. XII, 3, 3.

On this circumstance depends the easy motions of the fingers in performing music, and of the feet and arms in dancing and fencing, and of the hands in the use of tools in mechanic arts, as well as all the vital motions which animate and nourish organic bodies.

On the contrary, many animal motions by perpetual repetition are performed with less energy; as those who live near a waterfall, or a smith's forge, after a time, cease to hear them. And in those infectious diseases which are attended with fever, as the small-pox and measles, violent motions of the system are excited, which at length cease, and cannot again be produced by application of the same stimulating material; as when those are inoculated for the small-pox, who have before undergone that malady. Hence the repetition, which occasions animal actions for a time to be performed with greater energy, occasions them at length to become feeble, or to cease entirely.

To explain this difficult problem we must more minutely consider the catenations of animal motions, as described in Zoonomia, Vol. I, Sect. XVII. The vital motions, as suppose of the heart and arterial system, commence from the irritation occasioned by the stimulus of the blood, and then have this irritation assisted by the power of association; at the same time an agreeable sensation is produced by the due actions of the fibres, as in the secretions of the glands, which constitutes the pleasure of existence; this agreeable sensation is intermixed between every link of this diurnal chain of actions, and contributes to produce it by what is termed animal causation. But there is also a degree of the power of volition excited in consequence of this vital pleasure, which is also intermixed between the links of the chain of fibrous actions; and thus also contributes to its uniform easy and perpetual production.

The effects of surprise and novelty must now be considered by the patient reader, as they affect the catenations of action; and, I hope, the curiosity of the subject will excuse the prolixity of this account of it. When any violent stimulus breaks the passing current or catenation of our ideas, surprise is produced, which is accompanied with pain or pleasure, and consequent volition to examine the object of it, as explained in Zoonomia, Vol. I, Sect. XVIII, 17, and which never affects us in sleep. In our waking hours whenever an idea of imagination occurs, which is incongruous to our former experience, we feel another kind of surprise, and instantly dissever the train of imagination by the power of volition, and compare the incongruous idea with our previous knowledge of nature, and reject it by an act of reasoning, of which we are unconscious, termed in Zoonomia, "Intuitive Analogy," Vol. I, Sect. XVII, 7.

The novelty of any idea may be considered as affecting us with another kind of surprise, or incongruity, as it differs from the usual train of our ideas, and forms a new link in this perpetual chain; which, as it thus differs from the ordinary course of nature, we instantly examine by the voluntary efforts of intuitive analogy; or by reasoning, which we attend to; and compare it with the usual appearances of nature.

These ideas which affect us with surprise, or incongruity, or novelty, are attended with painful or pleasurable sensation; which we mentioned before as intermixing with all catenations of animal actions, and contributing to strengthen their perpetual and energetic production; and also exciting in some degree the power of volition, which also intermixes with the links of the chain of animal actions, and contributes to produce it.

Now by frequent repetition the surprise, incongruity, or novelty ceases; and, in consequence, the pleasure or pain which accompanied it, and also the degree of volition which was excited by that sensation of pain or pleasure; and thus the sensorial power of sensation and of volition are subducted from the catenation of vital actions, and they are in consequence produced much weaker, and at length cease entirely. Whence we learn why contagious matters induce their effects on the circulation but once; and why, in process of time, the vital movements are performed with less energy, and at length cease; whence the debilities of age, and consequent death.

VIII.
REPRODUCTION

But Reproduction with ethereal fires
New life rekindles, ere the first expires.
Canto II, line 13

I. The reproduction or generation of living organized bodies, is the great criterion or characteristic which distinguishes animation from mechanism. Fluids may circulate in hydraulic machines, or simply move in them, as mercury in the barometer or thermometer, but the power of producing an embryon which shall gradually acquire similitude to its parent, distinguishes artificial from natural organization.

The reproduction of plants and animals appears to be of two kinds, solitary and sexual; the former occurs in the formation of the buds of trees, and the bulbs of tulips; which for several successions generate other buds, and other bulbs, nearly similar to the parent, but constantly approaching to greater perfection, so as finally to produce sexual organs, or flowers, and consequent seeds.

The same occurs in some inferior kinds of animals; as the aphises in the spring and summer are viviparous for eight or nine generations, which successively produce living descendants without sexual intercourse, and are themselves, I suppose, without sex; at length in the autumn they propagate males and females, which copulate and lay eggs, which lie dormant during the winter, and are hatched by the vernal sun; while the truffle, and perhaps mushrooms amongst vegetables, and the polypus and taenia amongst insects, perpetually propagate themselves by solitary reproduction, and have not yet acquired male and female organs.

Philosophers have thought these viviparous aphides, and the taenia, and volvox, to be females; and have supposed them to have been impregnated long before their nativity within each other; so the taenia and volvox still continue to produce their offspring without sexual intercourse. One extremity of the taenia, is said by Linneus to grow old, whilst at the other end new ones are generated proceeding to infinity like the roots of grass. The Volvox globator is transparent, and carries within itself children and grandchildren to the fifth generation like the aphides; so that the taenia produces children and grandchildren longitudinally in a chain-like series, and the volvox propagates an offspring included within itself to the fifth generation; Systema Naturae.

Many microscopic animals, and some larger ones, as the hydra or polypus, are propagated by splitting or dividing; and some still larger animals, as oysters, and perhaps eels, have not yet acquired sexual organs, but produce a paternal progeny, which requires no mother to supply it with a nidus, or with nutriment and oxygenation; and, therefore, very accurately resemble the production of the buds of trees, and the wires of some herbaceous plants, as of knot-grass and of strawberries, and the bulbs of other plants, as of onions and potatoes; which is further treated of in Phytologia, Sect. VII.

The manner in which I suspect the solitary reproduction of the buds of trees to be effected, may also be applied to the solitary generation of the insects mentioned above, and probably of many others, perhaps of all the microscopic ones. It should be previously observed, that many insects are hermaphrodite, possessing both male and female organs of reproduction, as shell-snails and dew-worms; but that these are seen reciprocally to copulate with each other, and are believed not to be able to impregnate themselves; which belongs, therefore, to sexual generation, and not to the solitary reproduction of which I am now speaking.

As in the chemical production of any new combination of matter, two kinds of particles appear to be necessary; one of which must possess the power of attraction, and the other the aptitude to be attracted, as a magnet and a piece of iron; so in vegetable or animal combinations, whether for the purpose of nutrition or for reproduction, there must exist also two kinds of organic matter; one possessing the appetency to unite, and the other the propensity to be united. (See Zoonomia, octavo edition, Sect. XXXIX, 8.) Hence in the generation of the buds of trees, there are probably two kinds of glands, which acquire from the vegetable blood, and deposite beneath the cuticle of the tree two kinds of formative organic matter, which unite and form parts of the new vegetable embryon; which again uniting with other such organizations form the caudex, or the plumula, or the radicle, of a new vegetable bud.

A similar mode of reproduction by the secretion of two kinds of organic particles from the blood, and by depositing them either internally as in the vernal and summer aphis or volvox, or externally as in the polypus and taenia, probably obtains in those animals; which are thence propagated by the father only, not requiring a cradle, or nutriment, or oxygenation from a mother; and that the five generations, said to be seen in the transparent Volvox globator within each other, are perhaps the successive progeny to be delivered at different periods of time from the father, and erroneously supposed to be mothers impregnated before their nativity.

II. Sexual as well as solitary reproduction appears to be effected by two kinds of glands; one of which collects or secretes from the blood formative organic particles with appetencies to unite, and the other formative organic particles with propensities to be united. These probably undergo some change by a kind of digestion in their respective glands; but could not otherwise unite previously in the mass of blood from its perpetual motion.

The first mode of sexual reproduction seems to have been by the formation of males into hermaphrodites; that is, when the numerous formative glands, which existed in the caudex of the bud of a tree, or on the surface of a polypus, became so united as to form but two glands; which might then be called male and female organs. But they still collect and secrete their adapted particles from the same mass of blood as in snails and dew-worms, but do not seem to be so placed as to produce an embryon by the mixture of their secreted fluids, but to require the mutual assistance of two hermaphrodites for that purpose.

From this view-of the subject, it would appear that vegetables and animals were at first propagated by solitary generation, and afterwards by hermaphrodite sexual generation; because most vegetables possess at this day both male and female organs in the same flower, which Linneus has thence well called hermaphrodite flowers; and that this hermaphrodite mode of reproduction still exists in many insects, as in snails and worms; and, finally, because all the male quadrupeds, as well as men, possess at this day some remains of the female apparatus, as the breasts with nipples, which still at their nativity are said to be replete with a kind of milk, and the nipples swell on titillation.

Afterwards the sexes seem to have been formed in vegetables as in flowers, in addition to the power of solitary reproduction by buds. So in animals the aphis is propagated both by solitary reproduction as in spring, or by sexual generation as in autumn; then the vegetable sexes began to exist in separate plants, as in the classes monoecia and dioecia, or both of them in the same plant also, as in the class polygamia; but the larger and more perfect animals are now propagated by sexual reproduction only, which seems to have been the chef-d'oeuvre, or capital work of nature; as appears by the wonderful transformations of leaf-eating caterpillars into honey-eating moths and butterflies, apparently for the sole purpose of the formation of sexual organs, as in the silk-worm, which takes no food after its transformation, but propagates its species and dies.

III. *Recapitulation*

The microscopic productions of spontaneous vitality, and the next most inferior kinds of vegetables and animals, propagate by solitary generation only; as the buds and bulbs raised immediately from seeds, the lycoperdon tuber, with probably many other fungi, and the polypus, volvox, and taenia. Those of the next order propagate both by solitary and sexual reproduction, as those buds and bulbs which produce flowers as well as other buds or bulbs; and the aphis, and probably many other insects. Whence it appears, that many of those vegetables and animals, which are produced by solitary generation, gradually become more perfect, and at length produce a sexual progeny.

A third order of organic nature consists of hermaphrodite vegetables and animals, as in those flowers which have anthers and stigmas in the same corol; and in many insects, as leeches, snails, and worms; and perhaps all those reptiles which have no bones, according to the observation of M. Poupart, who thinks, that the number of hermaphrodite animals exceeds that of those which are divided into sexes; Mémoires de l'Académie des Sciences. These hermaphrodite insects I suspect to be incapable of impregnating themselves for reasons mentioned in Zoonomia, Sect. XXXIX, 6, 2.

And, lastly, the most perfect orders of animals are propagated by sexual intercourse only; which, however, does not extend to vegetables, as all those raised from seed produce some generations of buds or bulbs, previous to their producing flowers, as occurs not only in trees, but also in the annual plants. Thus three or four joints of wheat grow upon each other, before that which produces a flower; which joints are all separate plants growing over each other, like the buds of trees, previous to the uppermost; though this happens in a few months in annual plants, which requires as many years in the successive buds of trees; as is further explained in Phytologia, Sect. IX, 3, 1.

IV. *Conclusion*

Where climate is favourable, and salubrious food plentiful, there is reason to believe, that the races of animals perpetually improve by reproduction. The smallest microscopic animals become larger ones in a short time, probably by successive reproductions, as is so distinctly seen in the buds of seedling apple-trees, and in the bulbs of tulips raised from seed; both which die annually, and leave behind them one or many, which are more perfect than themselves, till they produce a sexual progeny, or flowers. To which may be added, the rapid improvement of our domesticated dogs, horses, rabbits, pigeons, which improve in size, or in swiftness, or in the sagacity of the sense of smell, or in colour, or other properties, by sexual reproduction.

The great Linneus having perceived the changes produced in the vegetable world by sexual reproduction, has supposed that not more than about sixty plants were at first created, and that all the others have been formed by their solitary or sexual reproductions; and adds, Suadent haec Creatoris leges a simplicibus ad composita ["From our Creator these laws from the simple to the complex"]; Genera Plantarum, preface to the Natural Orders, and Amoenitates Academicae, Vol. VI, 279.

This mode of reasoning may be extended to the most simple productions of spontaneous vitality.

There is one curious circumstance of animal life analogous in some degree to this wonderful power of reproduction; which is seen in the propagation of some contagious diseases. Thus one grain of variolous matter, inserted by inoculation, shall in about seven days stimulate the system into unnatural action; which in about seven days more produces ten thousand times the quantity of a similar material thrown out on the skin in pustules!

The mystery of reproduction, which alone distinguishes organic life from mechanic or chemic action, is yet wrapt in darkness. During the decomposition of organic bodies, where there exists a due degree of warmth with moisture, new microscopic animals of the most minute kind are produced; and these possess the wonderful power of reproduction, or of producing animals similar to themselves in their general structure, but with frequent additional improvements; which the preceding parent might in some measure have acquired by his habits of life or accidental situation.

But it may appear too bold in the present state of our knowledge on this subject, to suppose that all vegetables and animals now existing were originally derived from the smallest microscopic ones, formed by spontaneous vitality? and that they have by innumerable reproductions, during innumerable centuries of time, gradually acquired the size, strength, and excellence of form and faculties, which they now possess? and that such amazing powers were originally impressed on matter and spirit by the great Parent of Parents! Cause of Causes! Ens Entium!

IX.
STORGE

And Heaven-born Storge weaves the social chain.
Canto II, line 92

The Greek word Storge is used for the affection of parents to children; which was also visibly represented by the Stork or Pelican feeding her young with blood taken from her own wounded bosom. A number of Pelicans form a semicircle in shallow parts of the sea near the coast, standing on their long legs; and thus including a shoal of small fish, they gradually approach the shore; and seizing the fish as they advance, receive them into a pouch under their throats; and bringing them to land regurgitate them for the use of their young, or for their future support. Adanson, Voyage to Senegal. In this country the parent Pigeons both male and female swallow the grain or other seeds, which they collect for their young, and bring it up mixed with a kind of milk from their stomachs, with their bills inserted into the mouths of the young doves. J. Hunter's works.

 The affection of the parent to the young in experienced mothers may be in part owing to their having been relieved by them from the burden of their milk; but it is difficult to understand, how this affection commences in those mothers of the bestial world, who have not experienced this relief from the sucking of their offspring; and still more so to understand how female birds were at first induced to incubate their eggs for many weeks; and lastly how caterpillars, as of the silk-worm, are induced to cover themselves with a well-woven house of silk before their transformation.

These as well as many other animal facts, which are difficult to account for, have been referred to an inexplicable instinct; which is supposed to preclude any further investigation: but as animals seem to have undergone great changes, as well as the inanimate parts of the earth, and are probably still in a state of gradual improvement; it is not unreasonable to conclude, that some of these actions both of large animals and of insects, may have been acquired in a state preceding their present one; and have been derived from the parents to their offspring by imitation, or other kind of tradition; thus the eggs of the crocodile are at this day hatched by the warmth of the sun in Egypt; and the eggs of innumerable insects, and the spawn of fish, and of frogs, in this climate are hatched by the vernal warmth: this might be the case of birds in warm climates, in their early state of existence; and experience might have taught them to incubate their eggs, as they became more perfect animals, or removed themselves into colder climates: thus the ostrich is said to sit upon its eggs only in the night in warm situations, and both day and night in colder ones.

This love of the mother in quadrupeds to the offspring, whom she licks and cleans, is so allied to the pleasure of the taste or palate, that nature seems to have had a great escape in the parent quadruped not devouring her offspring. Bitches, and cats, and sows, eat the placenta; and if a dead offspring occurs, I am told, that also is sometimes eaten, and yet the living offspring is spared; and by that nice distinction the progenies of those animals are saved from destruction!

"Certior factus sum a viro rebus antiquissimis docto, quod legitur in Berosi operibus homines ante diluvium mulierum puerperarum placentam edidisse, quasi cibum delicatum in epulis luxuriosis; et quod hoc nefandissimo crimine movebatur Deus diluvio submergere terrarum incolas." [Something like "The author Berosso, one of the most learned of ancient men, has informed me that before the Flood the human placenta was held as luxurious food, and that this crime made God so angry that he drenched the inhabitants of the land."] Anon.

It may be finally concluded, that this affection from the parent to the progeny existed before animals were divided into sexes, and produced the beginning of sympathetic society, the source of which may perhaps be thus well accounted for; whenever the glandular system is stimulated into greater natural action within certain limits, an addition of pleasure is produced along with the increased secretion; this pleasure arising from the activity of the system is supposed to constitute the happiness of existence, in contradistinction to the ennui or taedium vitae; as shown in Zoonomia, Sect. XXXIII, 1.

Hence the secretion of nutritious juices occasioned by the stimulus of an embryon or egg in the womb gives pleasure to the parent for a length of time; whence by association a similar pleasure may be occasioned to the parent by seeing and touching the egg or fetus after its birth; and in lactescent animals an additional pleasure is produced by the new secretion of milk, as well as by its emission into the sucking lips of the infant. This appears to be one of the great secrets of Nature, one of those fine, almost invisible cords, which have bound one animal to another.

The females of lactiferous animals have thus a passion or inlet of pleasure in their systems more than the males, from their power of giving suck to their offspring; the want of the object of this passion, either owing to the death of the progeny, or to the unnatural fashion of their situation in life, not only deprives them of this innocent and virtuous source of pleasure; but has occasioned diseases, which have been fatal to many of them.

X.
EVE FROM ADAM'S RIB

Form'd a new sex, the mother of mankind.
Canto II, line 140

The mosaic history of Paradise and of Adam and Eve has been thought by some to be a sacred allegory, designed to teach obedience to divine commands, and to account for the origin of evil, like Jotham's fable of the trees; Judges IX, 8, or Nathan's fable of the poor man and his lamb; 2 Samuel XII, 1, or like the parables in the New Testament; as otherwise knowledge could not be said to grow upon one tree, and life upon another, or a serpent to converse; and lastly that this account originated with the magi or philosophers of Egypt, with whom Moses was educated, and that this part of the history, where Eve is said to have been made from a rib of Adam might have been an hieroglyphic design of the Egyptian philosophers, showing their opinion that Mankind was originally of both sexes united, and was afterwards divided into males and females: an opinion in later times held by Plato, and I believe by Aristotle, and which must have arisen from profound inquiries into the original state of animal existence.

XI.
HEREDITARY DISEASES

The feeble births acquired diseases chase,
Till Death extinguish the degenerate race.
Canto II, line 165

As all the families both of plants and animals appear in a state of perpetual improvement or degeneracy, it becomes a subject of importance to detect the causes of these mutations.

The insects, which are not propagated by sexual intercourse, are so few or so small, that no observations have been made on their diseases; but hereditary diseases are believed more to affect the offspring of solitary than of sexual generation in respect to vegetables; as those fruit trees, which have for more than a century been propagated only by ingrafting, and not from seeds, have been observed by Mr. Knight to be at this time so liable to canker, as not to be worth cultivation. From the same cause I suspect the degeneracy of some potatoes and of some strawberries to have arisen; where the curled leaf has appeared in the former, and barren flowers in the latter.

This may arise from the progeny by solitary reproduction so much more exactly resembling the parent, as is well seen in grafted trees compared with seedling ones; the fruit of the former always resembling that of the parent tree, but not so of the latter. The grafted scion also accords with the branch of the tree from whence it was taken, in the time of its bearing fruit; for if a scion be taken from a bearing branch of a pear or apple tree, I believe, it will produce fruit even the next year, or that succeeding; that is, in the same time that it would have produced fruit, if it had continued growing on the parent tree; but if the parent pear or apple tree has been cut down or headed, and scions are then, taken from the young shoots of the stem, and ingrafted; I believe those grafted trees will continue to grow for ten or twelve years, before they bear fruit, almost as long as seedling trees, that is they will require as much time, as those new shoots from the lopped trunk would require, before they produce fruit. It should thence be inquired, when grafted fruit trees are purchased, whether the scions were taken from bearing branches, or from the young shoots of a lopped trunk; as the latter, I believe, are generally sold, as they appear stronger plants. This greater similitude of the progeny to the parent in solitary reproduction must certainly make them more liable to hereditary diseases, if such have been acquired by the parent from unfriendly climate or bad nourishment, or accidental injury.

In respect to the sexual progeny of vegetables it has long been thought, that a change of seed or of situation is in process of time necessary to prevent their degeneracy; but it is now believed, that it is only changing for seed of a superior quality, that will better the product. At the same time it may be probably useful occasionally to intermix seeds from different situations together; as the anther-dust is liable to pass from one plant to another in its vicinity; and by these means the new seeds or plants may be amended, like the marriages of animals into different families.

As the sexual progeny of vegetables are thus less liable to hereditary diseases than the solitary progenies; so it is reasonable to conclude, that the sexual progenies of animals may be less liable to hereditary diseases, if the marriages are into different families, than if into the same family; this has long been supposed to be true, by those who breed animals for sale; since if the male and female be of different temperaments, as these are extremes of the animal system, they may counteract each other; and certainly where both parents are of families, which are afflicted with the same hereditary disease, it is more likely to descend to their posterity.

The hereditary diseases of this country have many of them been the consequence of drinking much fermented or spirituous liquor; as the gout always, most kinds of dropsy, and, I believe, epilepsy, and insanity. But another material, which is liable to produce diseases in its immoderate use, I believe to be common salt; the sea-scurvy is evidently caused by it in long voyages; and I suspect the scrofula, and consumption, to arise in the young progeny from the debility of the lymphatic and venous absorption produced in the parent by this innutritious fossile stimulus. The petechiae and vibices in the sea-scurvy and occasional haemorrhages evince the defect of venous absorption; the occasional haemoptoe at the commencement of pulmonary consumption, seems also to arise from defect of venous absorption; and the scrofula, which arises from the inactivity of the lymphatic absorbent system, frequently exists along with pulmonary as well as with mesenteric consumption. A tendency to these diseases is certainly hereditary, though perhaps not the diseases themselves; thus a less quantity of ale, cyder, wine, or spirit, will induce the gout and dropsy in those constitutions, whose parents have been intemperate in the use of those liquors; as I have more than once had occasion to observe.

Finally the art to improve the sexual progeny of either vegetables or animals must consist in choosing the most perfect of both sexes, that is the most beautiful in respect to the body, and the most ingenious in respect to the mind; but where one sex is given, whether male or female, to improve a progeny from that person may consist in choosing a partner of a contrary temperament.

As many families become gradually extinct by hereditary diseases, as by scrofula, consumption, epilepsy, mania, it is often hazardous to marry an heiress, as she is not unfrequently the last of a diseased family.

XII.
CHEMICAL THEORY OF ELECTRICITY AND MAGNETISM

> Then mark how two electric streams conspire
> To form the resinous and vitreous fire.
> Canto III, line 21

I. *Of Attraction and Repulsion.*

The motions, which accomplish the combinations and decompositions of bodies, depend on the peculiar attractions and repulsions of the particles of those bodies, or of the sides and angles of them; while the motions of the sun and planets, of the air and ocean, and of all bodies approaching to a general centre or retreating from it, depend on the general attraction or repulsion of those masses of matter. The peculiar attractions above mentioned are termed chemical affinities, and the general attraction is termed gravitation; but the peculiar repulsions of the particles of bodies, or the general repulsion of the masses of matter, have obtained no specific names, nor have been sufficiently considered; though they appear to be as powerful agents as the attractions.

The motions of ethereal fluids, as of magnetism and electricity, are yet imperfectly understood, and seem to depend both on chemical affinity, and on gravitation; and also on the peculiar repulsions of the particles of bodies, and on the general repulsion of the masses of matter.

In what manner attraction and repulsion are produced has not yet been attempted to be explained by modern philosophers; but as nothing can act, where it does not exist, all distant attraction of the particles of bodies, as well as general gravitation, must be ascribed to some still finer ethereal fluid; which fills up all space between the suns and their planets, as well as the interstices of coherent matter. Repulsion in the same manner must consist of some finer ethereal fluid; which at first projected the planets from the sun, and I suppose prevents their return to it; and which occasionally volatilizes or decomposes solid bodies into fluid or aerial ones, and perhaps into ethereal ones.

May not the ethereal matter which constitutes repulsion, be the same as the matter of heat in its diffused state; which in its quiescent state is combined with various bodies, as appears from many chemical explosions, in which so much heat is set at liberty? The ethereal matter, which constitutes attraction, we are less acquainted with; but it may also exist combined with bodies, as well as in its diffused state; since the specific gravities of some metallic mixtures are said not to accord with what ought to result from the combination of their specific gravities, which existed before their mixture; but their absolute gravities have not been attended to sufficiently; as these have always been supposed to depend on their quantity of matter, and situation in respect to the centre of the earth.

The ethereal fluids, which constitute peculiar repulsions and attractions, appear to gravitate round the particles of bodies mixed together; as those, which constitute the general repulsion or attraction, appear to gravitate round the greater masses of matter mixed together; but that which constitutes attraction seems to exist in a denser state next to the particles or masses of matter; and that which constitutes repulsion to exist more powerfully in a sphere further from them; whence many bodies attract at one distance, and repel at another. This may be observed by approaching to each other two electric atmospheres round insulated cork-balls; or by pressing globules of mercury, which roll on the surface, till they unite with it; or by pressing the drops of water,' which stand on a cabbage leaf, till they unite with it, and hence light is reflected from the surface of a mirror without touching it.

Thus the peculiar attractions and repulsions of the particles of bodies, and the general ones of the masses of matter, perpetually oppose and counteract each other; whence if the power of attraction should cease to act, all matter would be dissipated by the power of repulsion into boundless space; and if heat, or the power of repulsion, should cease to act, the whole world would become one solid mass, condensed into a point.

II. *Preliminary Propositions*

The following propositions concerning Electricity and Galvanism will either be proved by direct experiments, or will be rendered probable by their tending to explain or connect the variety of electric facts, to which they will be applied.

1. There are two kinds of electric ether, which exist either separately or in combination. That which is accumulated on the surface of smooth glass, when it is rubbed with a cushion, is here termed vitreous ether; and that which is accumulated on the surface of resin or sealing-wax, when it is rubbed with a cushion, is here termed resinous ether; and a combination of them, as in their usual state, may be termed neutral electric ethers.

2. Atmospheres of vitreous or of resinous or of neutral electricity surround all separate bodies, are attracted by them, and permeate those, which are called conductors, as metallic and aqueous and carbonic ones; but will not permeate those, which are termed nonconductors, as air, glass, silk, resin, sulphur.

3. The particles of vitreous electric ether strongly repel each other as they surround other bodies; but strongly attract the particles of resinous electric ether: in similar manner the particles of the resinous ether powerfully repel each other, and as powerfully attract those of the vitreous ether. Hence in their separate state they appear to occupy much greater space, as they, gravitate round insulated bodies, and are then only cognizable by our senses or experiments. They rush violently together through conducting substances, and then probably possess much less space in this their combined state. They thus resemble oxygen gas and nitrous gas; which rush violently together when in contact; and occupy less space when united, than either of them possessed separately before their union. When the two electric ethers thus unite, a chemical explosion occurs, like an ignited train of gunpowder; as they give out light and heat; and rend or fuse the bodies they occupy; which cannot be accounted for on the mechanical theory of Dr. Franklin.

4. Glass holds within it in combination much resinous electric ether, which constitutes a part of it, and which more forcibly attracts vitreous electric ether from surrounding bodies, which stands on it mixed with a less proportion of resinous ether like an atmosphere, but cannot unite with the resinous ether, which is combined with the glass; and resin, on the contrary, holds within it in combination much vitreous electric ether, which constitutes a part of it, and which more forcibly attracts resinous electric ether from surrounding bodies, which stands on it mixed with a less proportion of vitreous ether like an atmosphere, but cannot unite with the vitreous ether, which is combined with the resin.

As in the production of vitrification, those materials are necessary which contain much oxygen, as minium, and manganese; there is probably much oxygen combined with glass, which may thence be esteemed a solid acid, as water may be esteemed a fluid one. It is hence not improbable, that one kind of electric ether may also be combined with it, as it seems to affect the oxygen of water in the Galvanic experiments. The combination of the other kind of electric ether with wax or sulphur, is countenanced from those bodies, when heated or melted, being said to part with much electricity as they cool, and as it appears to affect the hydrogen in the decomposition of water by Galvanism.

5. Hence the nonconductors of electricity are of two kinds; such as are combined with vitreous ether, as resin, and sulphur; and such as are combined with resinous ether, as glass, air, silk. But both these kinds of nonconductors are impervious to either of the electric ethers; as those ethers being already combined with other bodies will not unite with each other, or be removed from their situations by each other. Whereas the perfect conducting bodies, as metals, water, charcoal, though surrounded with electric atmospheres, as they have neither of the electric ethers combined with them, suffer them to permeate and pass through them, whether separately or in their neutral state of reciprocal combination.

But it is probable, that imperfect conductors may possess more or less of either the vitreous or resinous ether combined with them, since their natural atmospheres are dissimilar as mentioned below; and that this makes them more or less imperfect conductors.

6. Those bodies which are perfect conductors, have probably neutral electric atmospheres gravitating round them consisting of an equal or

saturated mixture of the two electric ethers, whereas the atmospheres round the nonconducting bodies probably consist of an unequal mixture of the electric ethers, as more of the vitreous one round glass, and more of the resinous one round resin; and, it is probable, that these mixed atmospheres, which surround imperfect conducting bodies, consist also of different proportions of the vitreous and resinous ethers, according to their being more or less perfect conductors. These minute degrees of the difference of these electric atmospheres are evinced by Mr. Bennet's Doubler of Electricity, as shown in his work, and are termed by him Adhesive Electric Atmospheres, to distinguish them from those accumulated by art; thus the natural adhesive electricity of silver is more of the vitreous kind compared with that of zinc, which consists of a greater proportion of the resinous; that is, in his language, silver is positive and zinc negative. This experiment I have successfully repeated with Mr. Bennet's Doubler along with Mr. Swanwick.

7. Great accumulation or condensation of the separate electric ethers attract each other so strongly, that they will break a passage through nonconducting bodies, as through a plate of glass, or of air, and will rend bodies which are less perfect conductors, and give out light and heat like the explosion of a train of gunpowder; whence, when a strong electric shock is passed through a quire of paper, a bur, or elevation of the sheets, is seen on both sides of it occasioned by the explosion. Whence trees and stone walls are burst by lightning, and wires are fused, and inflammable bodies burnt, by the heat given out along with the flash of light, which cannot be explained by the mechanic theory.

8. When artificial or natural accumulations of these separate ethers are very minute in quantity or intensity, they pass slowly and with difficulty from one body to another, and require the best conductors for this purpose; whence many of the phenomena of the torpedo or gymnotus, and of Galvanism. Thus after having discharged a coated jar, if the communicating wire has been quickly withdrawn, a second small shock may be taken after the principal discharge, and this repeatedly two or three times.

Hence the charge of the Galvanic pile being very minute in quantity or intensity, will not readily pass through the dry cuticle of the hands, though it so easily passes through animal flesh or nerves, as this combination of charcoal with water seems to constitute the most perfect conductor yet known.

9. As light is reflected from the surface of a mirror before it actually touches it, and as drops of water are repelled from cabbage leaves without touching them, and as oil lies on water without touching it, and also as a fine needle may be made to lie on water without touching it, as shown by Mr. Melville in the Literary Essays of Edinburgh; there is reason to believe, that the vitreous and resinous electric ethers are repelled by, or will not pass through, the surfaces of glass or resin, to which they are applied. But though neither of these electric ethers passes through the surfaces of glass or resin, yet their attractive or repulsive powers pass through them: as the attractive or repulsive power of the magnet to iron passes through the atmosphere, and all other bodies which exist between them. So an insulated cork-ball, when electrised either with vitreous or resinous ether, repels another insulated cork-ball electrised with the same kind of ether, through half an inch of common air, though these electric atmospheres do not unite.

Whence it may be concluded, that the general attractive and repulsive ethers accompany the electric ethers as well as they accompany all other bodies; and that the electric ethers do not themselves attract or repel through glass or resin, as they cannot pass through them, but strongly attract each other when they come into contact, rush together, and produce an explosion of the sudden liberation of heat and light.

III. *Effect of Metallic Points*

1. When a pointed wire is presented by a person standing on the ground to an insulated conductor, on which either vitreous or resinous electricity is accumulated, the accumulated electricity will pass off at a much greater distance than if a metallic knob be fixed on the wire and presented in its stead.

2. The same occurs if the metallic point be fixed on the electrised conductor, and the finger of a person standing on the ground be presented to it, the accumulated electricity will pass off at a much greater distance, and indeed will soon discharge itself by communicating the accumulated electricity to the atmosphere.

3. If a metallic point be fixed on the prime conductor, and the flame of a candle be presented to it, on electrising the conductor either with vitreous or resinous ether, the flame of the candle is blown from the point, which must be owing to the electric fluid in its passage from the point carrying along with it a stream of atmospheric air.

The manner in which the accumulated electricity so readily passes off by a metallic point may be thus understood; when a metallic point stands erect from an electrised metallic plane, the accumulated electricity which exists on the extremity of the point, is attracted less than that on the other parts of the electrised surface. For the particle of electric matter immediately over the point is attracted by that point only, whereas the particles of electric matter over every other part of the electrised plane, is not only attracted by the parts of the plane immediately under them, but also laterally by the circumjacent parts of it; whence the accumulated electric fluid is pushed off at this point by that over the other parts being more strongly attracted to the plane.

Thus if a light insulated horizontal fly be constructed of wire with points fixed as tangents to the circle, it will revolve the way contrary to the direction of the points as long as it continues to be electrised. For the same reason as when a circle of cork, with a point of the cork standing from it like a tangent, is smeared with oil, and thrown upon a lake, it will continue to revolve backwards in respect to the direction of the point till all the oil is dispersed upon the lake, as first observed by Dr. Franklin; for the oil being attracted to all the other parts of the cork-circle more than towards the pointed tangent, that part over the point is pushed off and diffuses itself on the water, over which it passes without touching, and consequently without friction; and thus the cork revolves in the contrary direction.

As the flame of a candle is blown from a point fixed on an electrised conductor, whether vitreous or resinous electricity is accumulated on it, it shows that in both cases electricity passes from the point, which is a forcible argument against the mechanical theory of positive and negative electricity; because then the flame should be blown towards the point in one case, and from it in the other.

So the electric fly, as it turns horizontally, recedes from the direction of the points of the tangents, whether it be electrised with vitreous or resinous electricity; whereas if it was supposed to receive electricity, when electrised by resin, and to part with it when electrised by glass, it ought to revolve different ways; which also forcibly opposes the theory of positive and negative electricity.

As an electrised point with either kind of electricity causes a stream of air to pass from it in the direction of the point, it seems to affect the air much in the same manner as the fluid matter of heat affects it; that is, it will not readily pass through it, but will adhere to the particles of air, and is thus carried away with them.

From this it will also appear, that points do not attract electricity, properly speaking, but suffer it to depart from them; as it is there less attracted to the body which it surrounds, than by any other part of the surface.

And as a point presented to an electrised conductor facilitates the discharge of it, and blows the flame of a candle towards the conductor, whether vitreous or resinous electricity be accumulated upon it; it follows, that in both cases some electric matter passes from the point to the conductor, and that hence there are two electric ethers; and that they combine or explode when they meet together, and give out light and heat, and occupy less space in this their combined state, like the union of nitrous gas with oxygen gas.

IV. *Accumulation of Electric Ethers by Contact*

The electric ethers may be separately accumulated by contact of conductors with nonconductors, by vicinity of the two ethers, by heat, and by decomposition.

Glass is believed to consist in part of consolidated resinous ether, and thence to attract an electric atmosphere round it, which consists of a greater proportion of vitreous ether compared to the quantity of the resinous, as mentioned in Proposition No. 4. This atmosphere may stand off a line from the surface of the glass, though its attractive or repulsive power may extend to a much greater distance; and a more equally mixed electric atmosphere may stand off about the same distance from the surface of a cushion.

Now when a cushion is forcibly pressed upon the surface of a glass cylinder or plane, the atmosphere of the cushion is forced within that of the glass, and consequently the vitreous part of it is brought within the sphere of the attraction of the resinous ether combined with the glass, and therefore becomes attracted by it in addition to the vitreous part of the spontaneous atmosphere of the glass; and the resinous part of the atmosphere of the cushion is at the same time repelled by its vicinity to the combined resinous ether of the glass. From both which circumstances a vitreous ether alone surrounds the part of the glass on which the cushion is forcibly pressed; which does not, nevertheless, resemble an electrised coated jar; as this accumulation of vitreous ether on one side of the glass is not so violently condensed, or so forcibly attracted to the glass by the loose resinous ether on the other side of it, as occurs in the charged coated jar.

Hence as weak differences of the kinds or quantities of electricity do not very rapidly change place, if the cushion be suddenly withdrawn, with or without friction, I suppose an accumulation of vitreous electric ether will be left on the surface of the glass, which will diffuse itself on an insulated conductor by the assistance of points, or will gradually be dissipated in the air, probably like odours by the repulsion of its own particles, or may be conducted away by the surrounding air as it is repelled from it, or by the moisture or other impurities of the atmosphere. And hence I do not suppose the friction of the glass-globe to be necessary, except for the purpose of more easily removing the parts of the surface from the pressure of the cushion to the points of the prime conductor, and to bring them more easily into reciprocal contact.

When sealing wax or sulphur is rubbed by a cushion, exactly the same circumstance occurs, but with the different ethers; as the resinous ether of the spontaneous atmosphere of the cushion, when it is pressed within the spontaneous atmosphere of the sealing wax, is attracted by the solid vitreous ether, which is combined with it; and at the same time the vitreous ether of the cushion is repelled by it; and hence an atmosphere of resinous ether alone exists between the sealing wax and the cushion thus pressed together. It is nevertheless possible, that friction on both sealing wax and glass may add some facility to the accumulations of their opposite ethers by the warmth which it occasions. As most electric machines succeed best after being warmed, I think even in dry frosty seasons.

Though when a cushion is applied to a smooth surfaced glass, so as to intermix their electric atmospheres, the vitreous ether of the cushion is attracted by the resinous ether combined with the glass; but does not intermix with it, but only adheres to it: and as the glass turns round, the vitreous electric atmosphere stands on the solid resinous electric ether combined with the glass; and is taken away by the metallic points of the prime conductor.

Yet if the surface of the glass be roughened by scratching it with a diamond or with hard sand, a new event occurs; which is, that the vitreous ether attracted from the cushion by the resinous ether combined with the glass becomes adhesive to it; and stands upon the roughened glass, and will not quit the glass to go to the prime conductor; whence the surface of the glass having a vitreous electric atmosphere united, as it were, to its inequalities, becomes similar to resin; and will now attract resinous electric ether, like a stick of sealing wax, without combining with it. Whence this curious and otherwise unintelligible phenomenon, that smooth surfaced glass will give vitreous electric ether to an insulated conductor, and glass with a roughened surface will give resinous ether to it.

V. *Accumulation of electric ethers by vicinity*

Though the contact of a cushion on the whirling glass is the easiest method yet in use for the accumulation of the vitreous electric ether on an insulated conductor; yet there are other methods of effecting this, as by the vicinity of the two electric ethers with a nonconductor between them.

Thus I believe a great quantity of both vitreous and resinous electric ether may be accumulated in the following manner. Let a glass jar be coated within in the usual manner; but let it have a loose external coating, which can easily be withdrawn by an insulating handle. Then charge the jar, as highly as it may be, by throwing into it vitreous electric ether; and in this state hermetically seal it, if practicable, otherwise close it with a glass stopple and wax. When the external coating is drawn off by an insulating handle, having previously had a communication with the earth, it will possess an accumulation of resinous electric ether; and then touching it with your finger, a spark will be seen, and there will cease to be any accumulated ether.

Thus by alternately replacing this loose coating, and withdrawing it from the sealed charged jar, by means of an insulating handle; and by applying it to one insulated conductor, when it is in the vicinity of the jar; and to another insulated conductor, when it is withdrawn; vitreous electric ether may be accumulated on one of them, and resinous on the other; and thus I suspect an immense quantity of both ethers may be produced without friction or much labour, if a large electric battery was so contrived; and that it might be applied to many mechanical purposes, where other explosions are now used, as in the place of steam engines, or to rend rocks, or timber, or destroy invading armies!

The principle of this mode of accumulating the two electric ethers in some measure resembles that of Volta's Electrophorus and Bennet's Doubler.

VI. *Accumulation of electric ethers by heat and by decomposition*

When glass or amber is heated by the fire in a dry season, I suspect that it becomes in some degree electric; as either of the electric ethers which is combined with them may have its combination with those materials loosened by the application of heat; and that on this account they may more forcibly attract the opposite one from the air in their vicinity.

It has long been known, that a siliceous stone called the tourmalin, when its surfaces are polished, if it be laid down before the fire, will become electrified with vitreous, or what is called positive electricity on its upper surface; and resinous, or what is called negative electricity on its under surface; which I suppose lay in contact with somewhat which supported it near the fire.

In this experiment I suppose the tourmalin to be naturally combined with resinous electric ether like glass; which on one side next towards the fire by the increase of its attractive power, owing to the heat having loosened its combination with the earth of the stone, more strongly attracts vitreous electric ether from the atmosphere; which now stands on its surface: and then as the lower surface of the stone lies in contact with the hearth, the less quantity of vitreous ether is there repelled by the greater quantity of it on the upper surface; while the resinous ether is attracted by it: and the stone is thus charged like a coated jar with vitreous electric ether condensed on one side of it, and resinous on the other.

So cats, as they lie by the fire in a frosty day, become so electric as frequently to give a perceptible spark to one's finger from their ears without friction.

A fourth method of separating the two ethers would seem to be by the decomposition of metallic bodies, as in the experiment with Volta's Galvanic pile; which is said by Mr. Davy to act so much more powerfully, when an acid is added to the water used in the experiment; as will be spoken of below.

From experiments made by M. Saussure on the electricity of evaporated water from hot metallic vessels, and from those of china and glass, he found when the vessel was calcined or made rusty by the evaporating water, that the electricity of it was positive (or vitreous), and that from china or glass was negative (or resinous), Encyclopaedia Britannica, Art. Electricity, No. 206, which seems also to show, that vitreous electric ether was given out or produced by the corrosion of metals, and resinous ether from the evaporation of water.

VII. *The spark from the conductor, and of electric light*

When either the vitreous or resinous electric ether is accumulated on an insulated conductor, and an uninsulated conductor, as the finger of an attendant, is applied nearly in contact with it, what happens? The attractive and repulsive powers of the accumulated electric ether pass through the nonconducting plate of air, and if it be of the vitreous kind, it attracts the resinous electric ether of the finger towards it, and repels the vitreous electric ether of the finger from it.

Hence there exists for an instant a charged plate of air between the finger and the prime conductor, with an accumulation of vitreous ether on one side of it, and of resinous ether on the other side of it; and lastly these two kinds of electric ethers suddenly unite by their powerful attraction of each other, explode, and give out heat and light, and rupture the plate of nonconducting air, which separated them.

The rupture or disjunction of the plate of air is known by the sound of the spark, as of thunder; which shows that a vacuum of air was previously produced by the explosion of the electric fluids, and a vibration of the air in consequence of the sudden joining again of the sides of the vacuum.

The light which attends electric sparks and shocks, is not accounted for by the Theory of Dr. Franklin. I suspect that it is owing to the combination of the two electric ethers, from which as from all chemical explosions both light and heat are set at liberty, and because a smell is said to be perceptible from electric sparks, and even a taste which must be deduced from new combinations, or decompositions, as in other explosions: add to this that the same thing occurs, when electric shocks are passed through eggs in the dark, or through water, a luminous line is seen like the explosion of a train of gunpowder; lastly, whether light is really produced in the passage of the Galvanic electricity through the eyes, or that the sensation alone of light is perceived by its stimulating the optic nerve, has not yet been investigated; but I suspect the former, as it emits light from its explosion even in passing through eggs and through water, as mentioned above.

VIII. *The shock from the coated jar, and of electric condensation*

1. When a glass jar is coated on both sides, and either vitreous or resinous electricity is thrown upon the coating on one side, and there is a communication to the earth from the other side, the same thing happens as in the plate of air between the finger and prime conductor above described; that is, the accumulated electricity, if it be of the vitreous kind, on one coating of the glass jar will attract the resinous part of the electricity, which surrounds or penetrates the coating on the other side of the jar, and also repel the vitreous part of it; but this occurs on a much more extensive surface than in the instance of the plate of air between the finger and prime conductor.

The difference between electric sparks and shocks consists in this circumstance, that in the former the insulating medium, whether of air, or of thin glass, is ruptured in one part, and thus a communication is made between the vitreous and resinous ethers, and they unite immediately, like globules of quicksilver, when pressed forcibly together: but in the electric shock a communication is made by some conducting body applied to the other extremities of the vitreous, and of the resinous atmospheres, through which they pass and unite, whether both sides of the coated jar are insulated, or only one side of it.

And in this line, as they reciprocally meet, they appear to explode and give out light and heat, and a new combination of the two ethers is produced, as a residuum after the explosion, which probably occupies much less space than either the vitreous or resinous ethers did separately before. At the same time there may be another unrestrainable ethereal fluid yet unobserved, given out from this explosion, which rends oak trees, bursts stone-walls, lights inflammable substances, and fuses metals, or dissipates them in a calciform smoak, along with which great light and much heat are emitted, or these effects are produced by the heat and light only thus set at liberty by their synchronous and sudden evolution.

2. The curious circumstance of electric condensation appears from the violence of the shock of the coated jar compared with the strongest spark from an insulated conductor, though the latter possesses a much greater surface; when vitreous electric ether is thrown on one side of a coated jar, it attracts the resinous electric ether of the other side of the coated jar; and the same occurs, when resinous ether is thrown on one side of it, it attracts the vitreous ether of the other side of it, and thus the vitreous electric ether on one side of the jar, and the resinous ether on the other side of it become condensed, that is accumulated in less space, by their reciprocal attraction of each other.

This condensation of the two electric ethers owing to their reciprocal attraction appears from another curious event, that the thinner the glass jar is, the stronger will the charge be on the same quantity of surface, as then the two ethers approaching nearer without their intermixing attract each other stronger, and consequently condense each other more. And when the glass jar is very thin the reciprocal attractive powers of the vitreous and resinous ether attract each other so violently as at length to pass through the glass by rupturing it, in the same manner as a less forcible attraction of them ruptures and passes through the plate of air in the production of sparks from the prime conductor.

As these two ethers on each side of a charged coated jar so powerfully attract each other, when a communication is made between them by some conducting substance as in the common mode of discharging an electrised coated jar, they reciprocally pass to each other for the purpose of combining, as some chemical fluids are known to do; as when nitrous gas and oxygen gas are mixed together; whence as these fluids pass both ways to intermix with each other, and then explode; a bur appears on each side of a quire of paper well pressed together, when a strong electric shock is passed through it; which is occasioned by their explosion, like a train of gunpowder, and consequent emission of some other ethereal fluid, either those of heat and light or of some new one not yet observed. Whence it becomes difficult to explain, according to the theory of Dr. Franklin, which way the electric fluid passed; and which side of the coated jar contained positive and which the negative charge according to that doctrine.

But the theory of the ingenious Dr. Franklin failed also in explaining other phenomena of the coated jar; since if the positive electricity accumulated on one side of the jar repelled the electricity from the coating on the other side of it, so as to produce an electric vacuum; why should it be so eager, when a communication is made by some conducting body, to run into that vacuum by its attraction or gravitation, which has been made by its repulsion; as thus it seems to be violently attracted by the vacuum, from which it had previously repelled a fluid similar to itself, which is not easily to be comprehended.

3. There is another mode by which either vitreous or resinous electric ether is capable of condensation; which consists in contracting the volume, so as to diminish the surface of the electrised body; as was ingeniously shown by Dr. Franklin's experiment of electrising a silver tankard with a length of chain rolled up within it; and then drawing up the chain by a silk string, which weakened the electric attraction of the tankard; which was strengthened again by returning the chain into it; thus the condensation of an electrised cloud is believed to condense the electric ether, which it contains, and thus to occasion the lightning passing from one cloud to another, or from a cloud into the earth.

This experiment of the chain and tankard is said to succeed as well with what is termed negative electricity in the theory of Dr. Franklin, as with what is termed positive electricity; but in that theory the negative electricity means a less quantity or total deprivation or vacuity of that fluid; now to condense negative electricity by lowering the suspended chain into the tankard ought to make it less negative; whereas in this experiment I am told it becomes more so, as appears by its stronger repulsion of cork balls suspended on silk strings, and previously electrised by rubbed sealing wax: and if the negative electricity be believed to be a perfect vacuum of it, the condensation of a vacuum of electricity is totally incomprehensible; and this experiment alone seems to demonstrate the existence of two electric ethers.

IX. *Of Galvanic Electricity*

1. The conductors of electricity, as well as the nonconductors of it, have probably a portion of the vitreous and resinous ethers combined with them, and have also another portion of these ethers diffused round them, which forms their natural or spontaneous adhesive atmospheres; and which exists in different proportions round them correspondent in quantity to those which are combined with them, but opposite in kind.

These adhesive spontaneous atmospheres of electricity are shown to consist of different proportions or quantities of the electric ethers by Mr. Bennet's Doubler of Electricity, as mentioned in his work called New Experiments on Electricity, sold by Johnson. In this work, p. 91, the blade of a steel knife was evidently, in his language, positive, compared to a soft iron wire which was comparatively negative; so the adhesive electricity of gold, silver, copper, brass, bismuth, mercury, and various kinds of wood and stone, were what he terms positive or vitreous; and that of tin and zinc, what he terms negative or resinous.

Where these spontaneous atmospheres of diffused electricity surrounding two conducting bodies, as two pieces of silver, are perfectly similar, they probably do not intermix when brought into the vicinity of each other; but if these spontaneous atmospheres of diffused electricity are different in respect to the proportion of the two ethers, or perhaps in respect to their quantity, in however small degree either of these circumstances exists, they may be made to unite but with some difficulty; as the two metallic plates, suppose one of silver, and another of zinc, which they surround, must be brought into absolute or adhesive contact; or otherwise these atmospheres may be forced together so as to be much flattened, and compress each other where they meet, like small globules of quicksilver when pressed together, but without uniting.

This curious phenomenon may be seen in more dense electric atmospheres accumulated by art, as in the following experiment ascribed to Mr. Canton. Lay a wooden skewer the size of a goose-quill across a dry wine-glass, and another across another wine-glass; let the ends of them touch each other, as they lie in a horizontal line; call them X and Y; approach a rubbed glass-tube near the external end of the skewer X, but not so as to touch it; then separate the two skewers by removing the wine-glasses further from each other; and lastly, withdraw the rubbed glass-tube, and the skewer X will now be found to possess resinous electricity, which has been generally called negative or minus electricity; and the skewer Y will be found to possess vitreous, or what is generally termed positive or plus electricity.

The same phenomenon will occur if rubbed sealing wax be applied near to, but not in contact with, the skewer X, as the skewer X will then be left with an atmosphere of vitreous ether, and the skewer Y with one of resinous ether. These experiments also evince the existence of two electric fluids, as they cannot be understood from an idea of one being a greater or less quantity of the same material; as a vacuum of electric ether, brought near to one end of the skewer, cannot be conceived so to attract the ether as to produce a vacuum at the other end.

In this experiment the electric atmospheres, which are nearly of similar kinds, do not seem to touch, as there may remain a thin plate of air between them, in the same manner as small globules of mercury may be pressed together so as to compress each other, long before they intermix; or as plates of lead or brass require strongly to be pressed together before they acquire the attraction of cohesion; that is, before they come into real contact.

2. It is probable, that all bodies are more or less perfect conductors, as they have less or more of either of the electric ethers combined with them; as mentioned in Preliminary Proposition, No. VI, as they may then less resist the passage of either of the ethers through them. Whence some conducting bodies admit the junction of these spontaneous electric atmospheres, in which the proportions or quantities of the two ethers are not very different, with greater facility than others.

Thus in the common experiments, where the vitreous or resinous ether is accumulated by art, metallic bodies have been esteemed the best conductors, and next to these water, and all other moist bodies; but it was lately discovered, that dry charcoal, recently burnt, was a more perfect conductor than metals; and it appears from the experiments discovered by Galvani, which have thence the name of Galvanism, that animal flesh, and particularly perhaps the nerves of animals, both which are composed of much carbon and water, are the most perfect conductors yet discovered; that is, that they give the least resistance to the junction of the spontaneous electric atmospheres, which exist round metallic bodies, and which differ very little in respect to the proportions of their vitreous and resinous ingredients.

Thus also, though where the accumulated electricities are dense, as in charging a coated glass-jar, the glass, which intervenes, may be of considerable thickness, and may still become charged by the stronger attraction of the secondary electric ethers; but where the spontaneous adhesive electric atmospheres are employed to charge plates of air, as in the Galvanic pile, or probably to charge thin animal membranes or cuticles, as perhaps in the shock given by the torpedo or gymnotus, it seems necessary that the intervening nonconducting plate must be extremely thin, that it may become charged by the weaker attraction of these small quantities or difference of the spontaneous electric atmospheres; and in this circumstance only, I suppose, the shocks from the Galvanic pile, and from the torpedo and gymnotus, differ from those of the coated jar.

3. When atmospheres of electricity, which do not differ much in the quantity or proportion of their vitreous and resinous ethers, approach each other, they are not easily or rapidly united; but the predominant vitreous or resinous ether of one of them repels the similar ether of the opposed atmosphere, and attracts the contrary kind of ether.

The slowness or difficulty with, which atmospheres, which differ but little in kind or in density, unite with each other, appears not only from the experiment of Mr. Canton above related, but also from the repeated smaller shocks, which may be taken from a charged coated jar after the first or principal discharge, if the conducting medium has not been quickly removed, as is also mentioned above.

Hence those atmospheres of either kind of electric matter, which differ but very little from each other in kind or quantity, require the most perfect conductors to cause them to unite. Thus it appears by Mr. Bennet's doubler, as mentioned in the Preliminary Proposition, No. VI, that the natural adhesive atmosphere round silver contains more vitreous electricity than that naturally round zinc; but when thin plates of these metals, each about an ounce in weight, are laid on each other, or moderately pressed together, their atmospheres do not unite. For metallic plates, which when laid on each other, do not adhere, cannot be said to be in real contact, of which their not adhering is a proof; and in consequence a thin plate of air, or of their own repulsive ethers exists between them.

Hence when two plates of zinc and silver are thus brought in to the vicinity of each other, the plate of air between them, as they are not in adhesive contact, becomes like a charged coated jar; and if these two metallic plates are touched by your dry hands, they do not unite their electricities, as the dry cuticle is not a sufficiently good conductor; but if one of the metals be put above, and another under the tongue, the saliva and moist mucous membrane, muscular fibres, and nerves, supply so good a conductor, that this very minute electric shock is produced, and a kind of pungent taste is perceived.

When a plate or pencil of silver is put between the upper lip and the gum, and a plate or pencil of zinc under the tongue, a sensation of light is perceived in the eyes, as often as the exterior extremities of these metals are brought into contact; which is owing in like manner to the discharge of a very minute electric shock, which would not have been produced but by the intervention of such good conductors as moist membranes, muscular fibres, and nerves.

In this situation, a sensation of light is produced in the eyes; which seems to show, that these ethers pass through nerves more easily, than through muscular flesh simply; since the passage of them through the retina of the eyes from the upper gum to the parts beneath the tongue is a more distant one, than would otherwise appear necessary. It is not so easy to give the sensation of light in the eyes by passing a small shock of artificially accumulated electricity through, the eyes (though this may, I believe, be done) because this artificial accumulated electricity, as it passes with greater velocity than the spontaneous accumulations of it, will readily permeate the muscles or other moist parts of animal bodies;

whereas the spontaneous accumulations of electricity seem to require the best of all conductors, as animal nerves, to facilitate their passage.

4. In the Galvanic pile of Volta this electric shock becomes so much increased, as to pass by less perfect conductors, and to give shocks to the arms of the conducting person, if the cuticle of his hands be moistened, and even to show sparks like the coated jar; which appears to be effected in this manner. When a plate of silver is laid horizontally on a plate of zinc, the plate of air between them becomes charged like a coated jar; as the silver, naturally possessing more vitreous electric ether, repels the vitreous ether, which the zinc possesses in less quantity, and attracts the resinous ether of the zinc. Whence the inferior surface of the plate of zinc abounds now with vitreous ether, and its upper surface with resinous ether. Beneath this pair of plates lay a cloth moistened with water, or with some better conductor, as salt and water, or a slight acid mixed with water, or volatile alcali of ammoniac mixed with water, and this vitreous electric ether on the lower surface of the zinc plate will be given to the second silver plate which lies beneath it; and thus this second silver plate will possess not only its own natural vitreous atmosphere, which was denser or in greater quantity than that of the zinc plate next beneath it, but now acquires an addition of vitreous ether from the zinc plate above it, conducted to it through the moist cloth.

This then will repel more vitreous ether from the second zinc plate into the third silver one; and so on till the plates of air between the zincs and silvers are all charged, and each stronger and stronger, as they descend in the pile.

If the reader still prefers the Franklinian theory of positive and negative electricity, he will please to put the word positive for vitreous, and negative for resinous, and he will find the theory of the Galvanic pile equally thus accounted for.

5. When a Galvanic pile is thus placed, and a communication between the two ends of it is made by wires, so that the electric shocks pass through water, the water becomes decomposed in some measure, and oxygen is liberated from it at the point of one wire, and hydrogen at the point of the other; and this though a syphon of water be interposed between them. This curious circumstance seems to evince the existence of two electric ethers, which enter the water at different ends of the sy-

phon, and have chemical affinities to the component parts of it; the resinous ether sets at liberty the hydrogen at one end, and the vitreous ether the oxygen at the other end of the conducting medium.

Hence it must appear, that the longer the Galvanic pile, or the greater the number of the alternate pieces of silver and zinc that it consists of, the stronger will be the Galvanic shock; but there is another circumstance, difficult to explain, which is the perpetual decomposition of water by the Galvanic pile; when water is made the conducting medium between the two extremities of the pile.

As no conductors of electricity are absolutely perfect, there must be produced a certain accumulation of vitreous ether on one side of each charged plate of the Galvanic pile, and of resinous ether on the other side of it, before the discharge takes place, even though the conducting medium be in apparent contact. When the discharge does take place, the whole of the accumulated electricity explodes and vanishes; and then an instant of time is required for the silver and zinc again to attract from the air, or other bodies in their vicinity, their spontaneous natural atmospheres, and then another discharge ensues; and so repeatedly and perpetually till the surface of one of the metallic plates becomes so much oxydated or calcined, that it ceases to act.

Hence a perpetual motion may be said to be produced, with an incessant decomposition of water into the two gasses of oxygen and hydrogen; which must probably be constantly proceeding on all moist Surfaces, where a chain of electric conductors exists, surrounded with different proportions of the two electric ethers. Whence the ceaseless liberation of oxygen from the water has oxydated or calcined the ores of metals near the surface of the earth, as of manganese, of zinc into lapis calaminaris, of iron into various ochres, and other calciform ores. From this source also the corrosion of some metals may be traced, when they are immersed in water in the vicinity of each other, as when the copper sheathing of ships was held on by iron nails. And hence another great operation of nature is probably produced, I mean the restoration of oxygen to the atmosphere from the surface of the earth in dewy mornings, as well as from the perspiration of vegetable leaves; which atmospheric oxygen is hourly destructible by the respiration of animals and plants, by combustion, and by other oxydations.

6. The combination of the electric ethers with metallic bodies, before mentioned appears from the Galvanic pile; since, according to the experiments of Mr. Davy, when an acid is mixed with the water placed between the alternate pairs of silver and zinc plates, a much greater electric shock is produced by the same pile; and an anonymous writer in the Philosophical Magazine, No. 36, for May 1801, asserts, that when the intervening cloths or papers are moistened with pure alcali, as a solution of pure ammonia, the effect is greater than by any other material. It must here be observed, that both the acid and the alcaline solution, or common salt and water, and even water alone, in these experiments much erodes the plates of zinc, and somewhat tarnishes those of silver. Whence it would appear, that as by the repeated explosions of the two electric ethers in the conducting water, both oxygen and hydrogen are liberated; the oxygen erodes the zinc plates, and thus increases the Galvanic shock by liberating their combined electric ethers: and that this erosion is much increased by a mixture either of acid or of volatile alcali with the water. Further experiments are wanting on this subject to show whether metallic bodies emit either or both of the electric ethers at the time of their solution or erosion in acids or in alcalies.

X. *Of the two Magnetic Ethers*

1. Magnetism coincides with electricity in so many important points, that the existence of two magnetic ethers, as well as of two electric ones, becomes highly probable. We shall suppose, that in a common bar of iron or steel the two magnetic ethers exist intermixed or in their neutral state; which for the greater ease of speaking of them may be called arctic ether and antarctic ether; and in this state like the two electric fluids they are not cognizable by our senses of experiments.

When these two magnetic ethers are separated from each other, and the arctic ether is accumulated on one end of an iron or steel bar, which is then called the north pole of the magnet, and the antarctic ether is accumulated on the other end of the bar, and is then termed the south pole of the magnet; they become capable of attracting other pieces of iron or steel, and are thus cognizable by experiments.

It seems probable, that it is not the magnetic ether itself which attracts or repels particles of iron, but that an attractive and repulsive ether attends the magnetic ethers, as was shown to attend the electric ones in No. II, 9, of this Note; because magnetism does not pass through other bodies, as it does not escape from magnetised steel when in contact with other bodies; just as the electric fluids do not pass through glass, but the attractive and repellent ethers, which attend both the magnetic and electric ethers, pass through all bodies.

2. The prominent articles of analogical coincidence between magnetism and electricity are first, that when one end of an iron bar possesses an accumulation of arctic magnetic ether, or northern polarity; the other end possesses an accumulation of antarctic magnetic ether, or southern polarity; in the same manner as when vitreous electric ether is accumulated on one side of a coated glass jar, resinous electric ether becomes accumulated on the other side of it; as the vitreous and resinous ethers strongly attract each other, and strongly repel the ethers of the same denomination, but are prevented from intermixing by the glass plane between them; so the arctic and antarctic ethers attract each other, and repel those of similar denomination, but are prevented from intermixing by the iron or steel being a bad conductor of them; they will, nevertheless, sooner combine, when the bar is of soft iron, than when it is of hardened steel; and then they slowly combine without explosion, that is, without emitting heat and light like the electric ethers, and therefore resemble a mixture of oxygen and pure ammonia; which unite silently producing a neutral fluid without emitting any other fluids previously combined with them.

Secondly, If the north pole of a magnetic bar be approached near to the eye of a sewing needle, the arctic ether of the magnet attracts the antarctic ether, which resides in the needle towards the eye of it, and repels the arctic ether, which resides in the needle towards the point, precisely in the same manner as occurs in presenting an electrised, glass tube, or a rubbed stick of sealing wax to one extremity of two skewers insulated horizontally on wine-glasses in the experiment ascribed to Mr. Canton, and described in No. IX, 1, of this Additional Note, and also so exactly resembles the method of producing a separation and consequent accumulation of the two electric ethers by pressing a cushion on glass or on sealing wax, described in No. 4 of this Note, that their analogy is evidently apparent.

Thirdly, When much accumulated electricity is approached to one end of a long glass tube by a charged prime conductor, there will exist many divisions of the vitreous and resinous electricity alternately; as the vitreous ether attracts the resinous ether from a certain distance on the surface of the glass tube, and repels the vitreous ether; but, as this surface is a bad conductor, these reciprocal attractions and repulsions do not extend very far along it, but cease and recur in various parts of it. Exactly similar to this, when a magnetic bar is approximated to the end of a common bar of iron or steel, as described in Mr. Cavallo's valuable Treatise on Magnetism; the arctic ether of the north pole of the magnetic bar attracts the antarctic ether of the bar of common iron towards the end in contact, and repels the arctic ether; but, as iron and steel are as bad conductors of magnetism, as glass is of electricity, this accumulation of arctic ether extends but a little way, and then there exists an accumulation of antarctic ether; and thus reciprocally in three or four divisions of the bar, which now becomes magnetised, as the glass tube became electrised.

Another striking feature, which shows the sisterhood of electricity and magnetism, consists in the origin of both of them from the earth, or common mass of matter. The eduction of electricity from the earth is shown by an insulated cushion soon ceasing to supply either the vitreous or resinous ether to the whirling globe of glass or of sulphur; the eduction of magnetism from the earth appears from the following experiment: if a bar of iron be set upright on the earth in this part of the world, it becomes in a short time magnetical; the lower end possessing northern polarity, or arctic ether, and the higher end in consequence possessing southern polarity or antarctic ether; which may be well explained, if we suppose with Mr. Cavallo, that the earth itself is one great magnet, with its southern polarity or antarctic ether at the northern end of its axis; and, in consequence, that it attracts the arctic ether of the iron bar into that end of it which touches the earth, and repels the antarctic ether of the iron bar to the other end of it, exactly the same as when the southern pole of an artificial magnet is brought into contact with one end of a sewing needle.

3. The magnetic and electric ethers agree in the characters above mentioned, and perhaps in many others, but differ in the following ones. The electric ethers pass readily through metallic, aqueous, and carbonic bodies, but do not permeate vitreous or resinous ones; though on the surfaces of these they are capable of adhering, and of being accumulated by the approach or contact of other bodies; while the magnetic ethers will not permeate any bodies, and are capable of being accumulated only on iron and steel by the approach or contact of natural or artificial magnets, or of the earth; at the same time the attractive and repulsive powers both of the magnetic and electric ethers will act through all bodies, like those of gravitation and heat.

Secondly, The two electric ethers rush into combination, when they can approach each other, after having been separated and condensed, and produce a violent explosion emitting the heat and light, which were previously combined with them; whereas the two magnetic ethers slowly combine, after having been separated and accumulated on the opposite ends of a soft iron bar, and without emitting heat and light produce a neutral mixture, which, like the electric combination, ceases to be cognizable by our senses or experiments.

Thirdly, The wonderful property of the magnetic ethers, when separately accumulated on the ends of a needle, endeavouring to approach the two opposite poles of the earth; nothing similar to which has been observed in the electric ethers.

From these strict analogies between electricity and magnetism, we may conclude that the latter consists of two ethers as well as the former; and that they both, when separated by art or nature, combine by chemical affinity when they approach, the one exploding, and then consisting of a residuum after having emitted heat and light; and the other producing simply a neutralised fluid by their union.

XI. *Conclusion*

1. When two fluids are diffused together without undergoing any change of their chemical properties, they are said simply to be mixed, and not combined; as milk and water when poured together, or as oxygen and azote in the common atmosphere. So when salt or sugar is dif-

fused in water, it is termed solution, and not combination; as no change of their chemical properties succeeds.

But when an acid is mixed with a pure alcali a combination is produced, and the mixture is said to become neutral, as it does not possess the chemical properties which either of the two ingredients possessed in their separate state, and is therefore similar to neither of them. But when a carbonated alcali, as mild salt of tartar, is mixed with a mineral acid, they presently combine as above, but now the carbonic acid flies forcibly away in the form of gas; this, therefore, may be termed a kind of explosion, but cannot properly be so called, as the ethereal fluids of heat and light are not principally emitted, but an aerial one or gas; which may probably acquire a small quantity of heat from the combining matters.

But when strong acid of nitre is poured upon charcoal in fine powder, or upon oil of cloves, a violent explosion ensues, and the ethereal matters of heat and light are emitted in great abundance, and are dissipated; while in the former instance the oxygen of the nitrous acid unites with the carbone forming carbonic acid gas, and the azote escapes in its gaseous form; which may be termed a residuum after the explosion, and may be confined in a proper apparatus, which the heat and light cannot; for the former, if its production be great and sudden, bursts the vessels, or otherwise it passes slowly through them; and the latter passes through transparent bodies, and combines with opake ones.

But where ethers only are concerned in an explosion, as the two electric ones, which are previously difficult to confine in vessels; the repulsive ethers of heat and light are given out; and what remains is a combination of the two electric ethers; which in this state are attracted by all bodies, and form atmospheres round them.

These combined electric atmospheres must possess less heat and light after their explosion; which they seem afterwards to acquire at the time they are again separated from each other, probably from the combined heat and combined light of the cushion and glass, or of the cushion and resin; by the contact of which they are separated; and not from the diffused heat of them; but no experiments have yet been made to ascertain this fact, this combination of the vitreous and resinous ethers may be esteemed the residuum after their explosion.

2. Hence the essence of explosion consists in two bodies, which are

previously united with heat and light, so strongly attracting each other, as to set at liberty those two repulsive ethers; but it happens, that these explosive materials cannot generally be brought into each other's vicinity in a state of sufficient density; unless they are also previously combined with some other material beside the light and heat above spoken of: as in the nitrous acid, the oxygen is previously combined with azote; and is thus in a condensed state, before it is brought into the contact or vicinity of the carbone; there are however bodies which will slowly explode; or give out heat and light, without being previously combined with other bodies; as phosphorus in the common atmosphere, some dead fish in a certain degree of putridity, and some living insects probably by their respiration in transparent lungs, which is a kind of combustion.

But the two electric ethers are condensed by being brought into vicinity with each other with a nonconductor between them; and thus explode, violently as soon as they communicate, either by rupturing the interposed nonconductor, or by a metallic communication. This curious method of a previous condensation of the two exploding matters, without either of them being combined with any other material except with the ethers of heat and light, distinguishes, this ethereal explosion from that of most other bodies; and seems to have been the cause, which prevented the ingenious Dr. Franklin, and others since his time, from ascribing the powerful effects of the electric battery, and of lightning in bursting trees, inflaming combustible materials, and fusing metals, to chemical explosion; which it resembles in every other circumstance, but in the manner of the previous condensation of the materials, so as violently to attract each other, and suddenly set at liberty the heat and light, with which one or both of them were combined.

3. This combination of vitreous and resinous electric ethers is again destroyed or weakened by the attractions of other bodies; as they separate intirely, or exist in different proportions, forming atmospheres round conducting and nonconducting bodies; and in this they resemble other combinations of matters; as oxygen and azote, when united in the production of nitrous acid, are again separated by carbone; which attracts the oxygen more powerfully, than that attracts the azote, with which it is combined.

This mode of again separating the combined electric ethers by pressing them, as they surround bodies in different proportions, into each other's atmospheres, as by the glass and cushion, has not been observed respecting the decomposition of other bodies; when their minute particles are brought so near together as to decompose each other; which has thence probably contributed to prevent this decomposition of the two combined electric ethers from being ascribed to chemical laws; but, as far as we know, the attractive and repulsive atmospheres round the minute particles of bodies in chemical operations may act in a similar manner; as the attractive and repulsive atmospheres, which accompany the electric ethers surrounding the larger masses of matter, and that hence both the electric and the chemical explosions are subject to the same laws, and also the decomposition again of those particles, which were combined in the act of explosion.

4. It is probable that this theory of electric and magnetic attractions and repulsions, which so visibly exist in atmospheres round larger masses of matter, may be applied to explain the invisible attractions and repulsions of the minute particles of bodies in chemical combinations and decompositions, and also to give a clear idea of the attractions of the great masses of matter, which form the gravitations of the universe.

We are so accustomed to see bodies attract each other, when they are in absolute contact, as dew drops or particles of quicksilver forming themselves into spheres, as water rising in capillary tubes, the solution of salts and sugar in water, and the cohesion with which all hard bodies are held together, that we are not surprised at the attractions of bodies in contact with each other, but ascribe them to a law affecting all matter. In similar manner when two bodies in apparent contact repel each other, as oil thrown on water; or when heat converts ice into water and water into steam; or when one hard body in motion pushes another hard body out of its place; we feel no surprise, as these events so perpetually occur to us, but ascribe them as well as the attractions of bodies in contact with each other, to a general law of nature.

But when distant bodies appear to attract or repel each other, as we believe that nothing can act where it does not exist, we are struck with astonishment; which is owing to our not seeing the intermediate ethers, the existence of which is ascertained by the electric and magnetic facts above related.

From the facts and observations above mentioned electricity and magnetism consist each of them of two ethers, as the vitreous and resinous electric ethers, and the arctic and antarctic magnetic ethers. But as neither of the electric ethers will pass through glass or resin; and as neither of the magnetic ethers will pass through any bodies except iron; and yet the attractive and repulsive powers accompanying all these ethers permeate bodies of all kinds; it follows, that ethers more subtile than either the electric or magnetic ones attend those ethers forming atmospheres round them; as those electric and magnetic ethers themselves form atmospheres round other bodies.

This secondary atmosphere of the electric one appears to consist of two ethers, like the electric one which it surrounds: but these ethers are probably more subtile as they permeate all bodies; and when they unite by the reciprocal approach of the bodies, which they surround, they do not appear to emit heat and light, as the primary electric atmospheres do; and therefore they are simpler fluids, as they are not previously combined with heat and light. The secondary magnetic atmospheres are also probably more subtile or simple than the primary ones.

Hence we may suppose, that not only all the larger insulated masses of matter, but all the minute particles also, which constitute those masses, are surrounded by two ethereal fluids; which like the electric and magnetic ones attract each other forcibly, and as forcibly repel those of the same denomination; and at the same time strongly adhere to the bodies, which they surround. Secondly that these ethers are of the finer kind, like those secondary ones, which surround the primary electric and magnetic ethers; and that therefore they do not explode giving out heat and light when they unite, but simply combine, and become neutral; and lastly, that they surround different bodies in different proportions, as the vitreous and resinous electric ethers were shown to surround silver and zinc and many other metals in different proportions in No. IX, of this note.

5. For the greater ease of conversing on this subject, we shall call these two ethers, with which all bodies are surrounded, the masculine and the feminine ethers; and suppose them to possess the properties above mentioned. We should here however previously observe, that in chemical processes it is necessary, that the bodies, which are to combine

or unite with each other, should be in a fluid state, and the particles in contact with each other; thus when salt is dissolving in water, the particles of salt unite with those of the water, which touch them; these particles of water become saturated, and thence attract some of the saline particles with less force; which are therefore attracted from them by those behind; and the first particles of water are again saturated from the solid salt; or in some similar processes the saturated combinations may subside or evaporate, as in the union of the two electric ethers, or in the explosion of gunpowder, and thus those in their vicinity may approach each other. This necessity of a liquid form for the purpose of combination appears in the lighting of gunpowder, as well as in all other combustion, the spark of fire applied dissolves the sulphur, and liquifies the combined heat; and by these means a fluidity succeeds, and the consequent attractions and repulsions, which form the explosion.

The whole mixed mass of matter, of which the earth is composed, we suppose to be surrounded and penetrated by the two ethers, but with a greater proportion of the masculine ether than of the feminine. When a stone is elevated above the surface of the earth, we suppose it also to be surrounded with an atmosphere of the two ethers, but with a greater proportion of the feminine than of the masculine, and that these ethers adhere strongly by cohesion both to the earth and to the stone elevated above it. Now the greater quantity of the masculine ether of the earth becomes in contact with the greater quantity of the feminine ether of the stone above it; which it powerfully attracts, and at the same time repels the less quantity of the masculine ether of the stone. The reciprocal attractions of these two fluids, if not restrained by counter attractions, bring them together as in chemical combination, and thus they bring together the solid bodies, which they reciprocally adhere to; if they be not immovable; which solid bodies, when brought into contact, cohere by their own reciprocal attractions, and hence the mysterious affair of distant attraction or gravitation becomes intelligible, and consonant to the chemical combinations of fluids.

To further elucidate these various attractions, if the patient reader be not already tired, he will please to attend to the following experiment: let a bit of sponge suspended on a silk line be moistened with a solution of pure alcali, and another similar piece of sponge be moistened with a weak acid, and suspended near the former; electrize one of them with vitreous ether, and the other with resinous ether; as they hang with a thin plate of glass between them: now as these two electric ethers appear to attract each other without intermixing; as neither of them can pass through glass; they must be themselves surrounded with secondary ethers, which pass through the glass, and attract each other, as they become in contact; as these secondary ethers adhere to the primary vitreous and resinous ethers, these primary ones are drawn by them into each other's vicinity by the attraction of cohesion, and become condensed on each side of the glass plane; and then when the glass plane is withdrawn, the two electric ethers being now in contact rush violently together, and draw along with them the pieces of moistened sponge, to which they adhere; and finally the acid and alcaline liquids being now brought into contact combine by their chemical affinity.

The repulsions of distant bodies are also explicable by this idea of their being surrounded with two ethers, which we have termed masculine and feminine for the ease of conversing about them; and have compared them to vitreous and resinous electricity, and to arctic and antarctic magnetism. As when two particles of matter, or two larger masses of it, are surrounded both with their masculine ethers, these ethers repel each other or refuse to intermix; and in consequence the bodies to which they adhere, recede from each other; as two cork-balls suspended near each other, and electrised both with vitreous or both with resinous ether, repel each other; or as the extremities of two needles magnetised both with arctic, or both with antarctic ether, repel each other; or as oil and water surrounded both with their masculine, or both with their feminine ethers, repel each other without touching; so light is believed to be reflected from a mirror without touching its surface, and to be bent towards the edge of a knife, or refracted by its approach from a rarer medium into a denser one, by the repulsive ether of the mirror, and the attractive ones of the knife-edge, and of the denser medium. Thus a polished tea-cup slips on the polished saucer probably without their actual contact with each other, till a few drops of water are interposed between them by capillary attraction, and prevent its sliding by their tenacity. And so, lastly, one hard body in motion pushes another hard body out of its place by their repulsive ethers without being in contact;

as appears from their not adhering to each other, which all bodies in real contact are believed to do. Whence also may be inferred the reason why bodies have been supposed to repel at one distance and attract at another, because they attract when their particles are in contact with each other, and either attract or repel when at a distance by the intervention of their attractive or repulsive ethers.

Thus have I endeavoured to take one step further back into the mystery of the gravitation and repulsion of bodies, which appeared to be distant from each other, as of the sun and planets, as I before endeavoured to take one step further back into the mysteries of generation in my account of the production of the buds of vegetables in Phytologia. With what success these have been attended I now leave to the judgment of philosophical readers, from which I can make no appeal.

XIII.
ANALYSIS OF TASTE

Fond Fancy's eye recalls the form divine,
And Taste sits smiling upon Beauty's shrine.
Canto III, line 221

The word Taste in its extensive application may express the pleasures received by any of our senses, when excited into action by the stimulus of external objects; as when odours stimulate the nostrils, or flavours the palate; or when smoothness, or softness, are perceived by the touch, or warmth by its adapted organ of sense. The word Taste is also used to signify the pleasurable trains of ideas suggested by language, as in the compositions of poetry and oratory. But the pleasures, consequent to the exertions of our sense of vision only, are designed here to be treated of, with occasional references to those of the ear, when they elucidate each other.

When any of our organs of sense are excited into their due quantity of action, a pleasurable sensation succeeds, as shown in Zoonomia, Vol. I, Sect. IV. These are simply the pleasures attending perception, and not those which are termed the pleasures of Taste; which consist of additional pleasures arising from the peculiar forms or colours of objects, or of their peculiar combinations or successions, or from other agreeable trains of ideas previously associated with them.

There are four sources of pleasure attendant on the excitation of the nerves of vision by light and colours, besides that simply of perception above mentioned; the first is derived from a degree of novelty of the forms, colours, numbers, combinations, or successions, and visible objects. The second is derived from a degree of repetition of their forms, colours, numbers, combinations, or successions. Where these two circumstances exist united in certain quantities, and compose the principal part of a landscape, it is termed picturesque by modern writers. The third source of pleasure from the perception of the visible world may be termed the melody of colours, which will be shown to coincide with melody of sounds: this circumstance may also accompany the picturesque, and will add to the pleasure it affords. The fourth source of pleasure from the perception of visible objects is derived from the previous association of other pleasurable trains of ideas with certain forms, colours, combinations, or successions of them. Whence the beautiful, sublime, romantic, melancholic, and other emotions, which have not acquired names to express them. We may add, that all these four sources of pleasure from perceptions are equally applicable to those of sounds as of sights.

I. *Novelty or infrequency of visible objects*

The first circumstance, which suggests an additional pleasure in the contemplation of visible objects, besides that of simple perception, arises from their novelty or infrequency; that is from the unusual combinations or successions of their forms or colours. From this source is derived the perpetual cheerfulness of youth, and the want of it is liable to add a gloom to the countenance of age. It is this which produces variety in landscape compared with the common course of nature, an intricacy which incites investigation, and a curiosity which leads to explore the works of nature. Those who travel into foreign regions instigated by curiosity, or who examine and unfold the intricacies of sciences at home, are led by novelty; which not only supplies ornament to beauty or to grandeur, but adds agreeable surprise to the point of the epigram, and to the double meaning of the pun, and is courted alike by poets and philosophers.

It should be here premised, that the word Novelty, as used in these pages, admits of degrees or quantities, some objects, or the ideas excited by them, possessing more or less novelty, as they are more or less unusual. Which the reader will please to attend to, as we have used the word Infrequency of objects, or of the ideas excited by them, to express the degrees or quantities of their novelty.

The source, from which is derived the pleasure of novelty, is a metaphysical inquiry of great curiosity, and will on that account excuse my here introducing it. In our waking hours whenever an idea occurs, which is incongruous to our former experience, we instantly dissever the train of imagination by the power of volition; and compare the incongruous idea with our previous knowledge of nature, and reject it. This operation of the mind has not yet acquired a specific name, though it is exerted every minute of our waking hours, unless it may be termed Intuitive Analogy. It is an act of reasoning of which we are unconscious except by its effects in preserving the congruity of our ideas; Zoonomia, Vol. I, Sect. XVII, 5, 7.

In our sleep as the power of volition is suspended, and consequently that of reason, when any incongruous ideas occur in the trains of imagination, which compose our dreams; we cannot compare them with our previous knowledge of nature and reject them; whence arises the perpetual inconsistency of our sleeping trains of ideas; and whence in our dreams we never feel the sentiment of novelty; however different the ideas, which present themselves, may be from the usual course of nature.

But in our waking hours, whenever any object occurs which does not accord with the usual course of nature, we immediately and unconsciously exert our voluntary power, and examine it by intuitive analogy, comparing it with our previous knowledge of nature. This exertion of our volition excites many other ideas, and is attended with pleasurable sensation; which constitutes the sentiment of novelty. But when the object of novelty stimulates us so forcibly as suddenly to disunite our passing trains of ideas, as if a pistol be unexpectedly discharged, the emotion of surprise is experienced; which by exciting violent irritation and violent sensation, employs for a time the whole sensorial energy, and thus dissevers the passing trains of ideas; before the power of volition has time to compare them with the usual phenomena of nature; but as the painful emotion of fear is then generally added to that of surprise, as every one experiences, who hears a noise in the dark, which he cannot immediately account for; this great degree of novelty, when it produces much surprise, generally ceases to be pleasurable, and does not then belong to objects of taste.

In its less degree surprise is generally agreeable, as it simply expresses the sentiment occasioned by the novelty of our ideas; as in common language we say, we are agreeably surprised at the unexpected meeting with a friend, which not only expresses the sentiment of novelty, but also the pleasure from other agreeable ideas associated with the object of it.

It must appear from hence, that different persons must be affected more or less agreeably by different degrees or quantities of novelty in the objects of taste; according to their previous knowledge of nature, or their previous habits or opportunities of attending to the fine arts. Thus before its nativity the fetus experiences the perceptions of heat and cold, of hardness and softness, of motion and rest, with those perhaps of hunger and repletion, sleeping and waking, pain and pleasure; and perhaps some other perceptions, which may at this early time of its existence have occasioned perpetual trains of ideas. On its arrival into the world the perceptions of light and sound must by their novelty at first dissever its usual trains of ideas and occasion great surprise; which after a few repetitions will cease to be disagreeable, and only excite the emotion from novelty, which has not acquired a separate name, but is in reality a less degree of surprise; and by further experience the sentiment of novelty, or any degree of surprise, will cease to be excited by the sounds or sights, which at first excited perhaps a painful quantity of surprise.

It should here be observed, that as the pleasure of novelty is produced by the exertion of our voluntary power in comparing uncommon objects with those which are more usually exhibited; this sentiment of novelty is less perceived by those who do not readily use the faculty of volition, or who have little previous knowledge of nature, as by very ignorant or very stupid people, or by brute animals; and that therefore to be affected with this circumstance of the objects of Taste requires some previous knowledge of such kinds of objects, and some degree of mental exertion.

Hence when a greater variety of objects than usual is presented to the eye, or when some intricacy of forms, colours, or reciprocal locality more than usual accompanies them, it is termed novelty if it only excites the exertion of intuitive comparison with the usual order of nature, and affects us with pleasurable sensation; but is termed surprise, if it suddenly dissevers our accustomed habits of motion, and is then more generally attended with disagreeable sensation. To this circumstance attending objects of taste is to be referred what is termed wild and irregular in landscapes, in contradistinction to the repetition of parts or uniformity spoken of below. We may add, that novelty of notes and tones in music, or of their combinations or successions, are equally agreeable to the ear, as the novelty of forms and colours, and of their combinations or successions are to the eye; but that the greater quantity or degree of novelty, the sentiment of which is generally termed Surprise, is more frequently excited by unusual or unexpected sounds; which are liable to alarm us with fear, as well as surprise us with novelty.

II. *Repetition of visible objects*

The repeated excitement of the same or similar ideas with certain intervals of time, or distances of space between them, is attended with agreeable sensations, besides that simply of perception; and, though it appears to be diametrically opposite to the pleasure arising from the novelty of objects above treated of, enters into the compositions of all the agreeable arts.

The pleasure arising from the repetition of similar ideas with certain intervals of time or distances of space between them is a subject of great metaphysical curiosity, as well as the source of the pleasure derived from novelty, which will I hope excuse its introduction in this place.

The repetitions of motions may be at first produced either by volition, or by sensation, or by irritation, but they soon become easier to perform than any other kinds of action, because they soon become associated together; and thus their frequency of repetition, if as much sensorial power be produced during every reiteration, as is expended, adds to the facility of their production.

If a stimulus be repeated at uniform intervals of time, the action, whether of our muscles or organs of sense, is produced with still greater facility or energy; because the sensorial power of association, mentioned above, is combined with the sensorial power of irritation; that is in common language, the acquired habit assists the power of the stimulus.

This not only obtains in the annual, lunar, and diurnal catenations of animal motions, as explained in Zoonomia, Sect. XXXVI, which are thus performed with great facility and energy; but in every less circle of actions or ideas, as in the burden of a song, or the reiterations of a dance. To the facility and distinctness, with which we hear sounds at repeated intervals, we owe the pleasure, which we receive from musical time, and from poetic time, as described in Botanic Garden, Vol. II, Interlude III. And to this the pleasure we receive from the rhimes and alliterations of modern versification; the source of which without this key would be difficult to discover.

There is no variety of notes referable to the gamut in the beating of a drum, yet if it be performed in musical time, it is agreeable to our ears; and therefore this pleasurable sensation must be owing to the repetition of the divisions of the sounds at certain intervals of time, or musical bars. Whether these times or bars are distinguished by a pause, or by an emphasis, or accent, certain it is, that this distinction is perpetually repeated; otherwise the ear could not determine instantly, whether the successions of sound were in common or in triple time.

But besides these little circles of musical time, there are the greater returning periods, and the still more distinct choruses; which, like the rhimes at the end of verses, owe their beauty to repetition; that is, to the facility and distinctness with which we perceive sounds, which we expect to perceive or have perceived before; or in the language of this work, to the greater ease and energy with which our organ is excited by the combined sensorial powers of association and irritation, than by the latter singly.

This kind of pleasure arising from repetition, that is from the facility and distinctness with which we perceive and understand repeated sensations, enters into all the agreeable arts; and when it is carried to excess is termed formality. The art of dancing like that of music depends for a great part of the pleasure, it affords, on repetition; architecture, especially the Grecian, consists of one part being a repetition of another, and hence the beauty of the pyramidal outline in landscape-painting; where one side of the picture may be said in some measure to balance the other. So universally does repetition contribute to our pleasure in the fine arts, that beauty itself has been defined by some writers to consist in a due combination of uniformity and variety: Zoonomia, Vol. I, Sect. XXII, 2, 1.

Where these repetitions of form, and reiterations of colour, are produced in a picture or a natural landscape, in an agreeable quantity, it is termed simplicity, or unity of character; where the repetition principally is seen in the disposition or locality of the divisions, it is called symmetry, proportion, or grouping the separate parts; where this repetition is most conspicuous in the forms of visible objects, it is called regularity or uniformity; and where it affects the colouring principally, the artists call it breadth of colour.

There is nevertheless, an excess of the repetition of the same or similar ideas, which ceases to please, and must therefore be excluded from compositions of Taste in painted landscapes, or in ornamented gardens; which is then called formality, monotony, or insipidity. Why the excitation of ideas should give additional pleasure by the facility and distinctness of their production for a certain time, and then cease to give additional pleasure; and gradually to give less pleasure than that, which attends simple exertion of them; is another curious metaphysical problem, and deserves investigation.

In our waking hours a perpetual voluntary exertion, of which we are unconscious, attends all our new trains of ideas, whether those of imagination or of perception; which by comparing them with our former experience preserves the consistency of the former, by rejecting such as are incongruous; and adds to the credibility of the latter, by their analogy to objects of our previous knowledge: and this exertion is attended with pleasurable sensation. After very frequent repetition these trains of ideas do not excite the exertion of this intuitive analogy, and in consequence are not attended with additional pleasure to that simply of perception; and by continued repetition they at length lose even the pleasure simply of perception, and thence finally cease to be excited; whence one cause of the torpor of old age, and of death, as spoken of in Additional Note, No. VII, 3, of this work.

When there exists in any landscape a certain number and diversity of forms and colours, or of their combinations or successions, so as to produce a degree of novelty; and that with a certain repetition, or arrangement of parts, so as to render them gradually comprehensible or easily compared with the usual course of nature; if this agreeable combination of visible objects be on a moderate scale, in respect to magnitude, and form the principal part of the landscape, it is termed Picturesque by modern artists; and when such a combination of forms and colours contains many easy flowing curves and smooth surfaces, the delightful sentiment of Beauty becomes added to the pleasure of the Picturesque.

If the above agreeable combination of novelty and repetition exists on a larger scale with more projecting rocks, and deeper dells, and perhaps with a somewhat greater proportion of novelty than repetition, the landscape assumes the name of Romantic; and if some of these forms or combinations are much above the usual magnitude of similar objects, the more interesting sentiment of Sublimity becomes mixed with the pleasure of the romantic.

III. *Melody of Colours*

A third source of pleasure arising from the inspection of visible objects, besides that of simple perception, arises from what may be termed melody of colours, as certain colours are more agreeable, when they succeed each other; or when they are disposed in each other's vicinity, so as successively to affect the organ of vision.

In a paper on the colours seen in the eye after looking for some time on luminous objects, published by Dr. Darwin of Shrewsbury in the

Philosophical Transactions, Vol. 76, it is evidently shown, that we see certain colours not only with greater ease and distinctness, but with relief and pleasure, after having for some time inspected other certain colours; as green after red, or red after green; orange after blue, or blue after orange; yellow after violet, or violet after yellow; this, he shows, arises from the ocular spectrum of the colour last viewed coinciding with the irritation of the colour now under contemplation.

Thus if you make a dot with ink in the centre of a circle of red silk the size of a letter-wafer, and place it on a sheet of white paper, and look on it for a minute without moving your eyes; and then gently turn them on the white paper in its vicinity, or gently close them, and hold one hand an inch or two before them, to prevent too much light from passing through the eyelids, a circular spot of pale green will be seen on the white paper, or in the closed eye; which is called the ocular spectrum of the red silk, and is formed as Dr. Darwin shows by the pandiculation or stretching of the fine fibrils, which constitute the extremities of the optic nerve, in a direction contrary to that, in which they have been excited by previously looking at a luminous object, till they become fatigued; like the yawning or stretching of the larger muscles after acting long in one direction.

If at this time the eye, fatigued by looking long at the centre of the red silk, be turned on paper previously coloured with pale green; the circular spot or ocular spectrum will appear of a much darker green; as now the irritation from the pale green paper coincides with the pale green spectrum remaining in the eye, and thus excites those fibres of the retina into stronger action; on this account some colours are seen more distinctly, and consequently more agreeably after others; or when placed in the vicinity of others; thus if orange-coloured letters are painted on a blue ground, they may be read at as great distance as black on white, perhaps at a greater.

The colours, which are thus more distinct when seen in succession are called opposite colours by Sir Isaac Newton in his Optics, Book I, Part 2, and may be easily discovered by any one, by the method above described; that is by laying a coloured circle of paper or silk on a sheet of white paper, and inspecting it some time with steady eyes, and then either gently closing them, or removing them on another part of the white paper, and the ocular spectrum or opposite colour becomes visible in the eye.

Sir Isaac Newton has observed, that the breadths of the seven primary colours in the sun's image refracted by a prism, are proportioned to the seven musical notes of the gamut; or to the intervals of the eight sounds contained in an octave.

From this curious coincidence, it has been proposed to produce a luminous music, consisting of successions or combinations of colours, analogous to a tune in respect to the proportions above mentioned. This might be performed by a strong light, made by means of Mr. Argand's lamps, passing through coloured glasses, and falling on a defined part of the wall, with moveable blinds before them, which might communicate with the keys of a harpsichord, and thus produce at the same time visible and audible music in unison with each other.

Now as the pleasure we receive from the sensation of melodious notes, independent of musical time, and of the previous associations of agreeable ideas with them, must arise from our hearing some proportions of sounds after others more easily, distinctly, or agreeably; and as there is a coincidence between the proportions of the primary colours, and the primary sounds, if they may be so called; the same laws must probably govern the sensations of both. In this circumstance therefore consists the sisterhood of Music and Painting; and hence they claim a right to borrow metaphors from each other: musicians to speak of the brilliancy of sounds, and the light and shade of a concerto; and painters of the harmony of colours, and the tone of a picture.

This source of pleasure received from the melodious succession of colours or of sounds must not be confounded with the pleasure received from the repetition of them explained above, though the repetition, or division of musical notes into bars, so as to produce common or triple time, contributes much to the pleasure of music; but in viewing a fixed landscape nothing like musical time exists; and the pleasure received therefore from certain successions of colours must depend only on the more easy or distinct action of the retina in perceiving some colours after others, or in their vicinity, like the facility or even pleasure with which we act with contrary muscles in yawning or stretching after having been fatigued with a long previous exertion in the contrary direction.

Hence where colours are required to be distinct, those which are opposite to each other, should be brought into succession or vicinity; as red and green, orange and blue, yellow and violet; but where colours are required to intermix imperceptibly, or slide into each other, these should not be chosen; as they might by contrast appear too glaring or tawdry. These gradations and contrasts of colours have been practically employed both by the painters of landscape, and by the planters of ornamental gardens; though the theory of this part of the pleasure derived from visible objects was not explained before the publication of the paper on ocular spectra above mentioned; which is reprinted at the end of the first part of Zoonomia, and has thrown great light on the actions of the nerves of sense in consequence of the stimulus of external bodies.

IV. *Association of agreeable sentiments with visible objects*

Besides the pleasure experienced simply by the perception of visible objects, it has been already shown, that there is an additional pleasure arising from the inspection of those, which possess novelty, or some degree of it; a second additional pleasure from those, which possess in some degree a repetition of their parts; and a third from those, which possess a succession of particular colours, which either contrast or slide into each other, and which we have termed melody of colours.

We now step forward to the fourth source of the pleasures arising from the contemplation of visible objects besides that simply of perception, which consists in our previous association of some agreeable sentiment with certain forms or combinations of them. These four kinds of pleasure singly or in combination constitute what is generally understood by the word Taste in respect to the visible world; and by parity of reasoning it is probable, that the pleasurable ideas received by the other senses, or which are associated with language, may be traced to similar sources.

It has been shown by Bishop Berkeley in his ingenious essay on vision, that the eye only acquaints us with the perception of light and colours; and that our idea of the solidity of the bodies, which reflect them, is learnt by the organ of touch: he therefore calls our vision the language of touch, observing that certain gradations of the shades of colour, by our previous experience of having examined similar bodies by our hands or lips, suggest our ideas of solidity, and of the forms of solid bodies; as when we view a tree, it would otherwise appear to us a flat green surface, but by association of ideas we know it to be a cylindrical stem with round branches. This association of the ideas acquired by the sense of touch with those of vision, we do not allude to in the following observations, but to the agreeable trains or tribes of ideas and sentiments connected with certain kinds of visible objects.

V. *Sentiment of Beauty*

Of these catenations of sentiments with visible objects, the first is the sentiment of Beauty or Loveliness; which is suggested by easy-flowing curvatures of surface, with smoothness; as is so well illustrated in Mr. Burke's Essay on the Sublime and Beautiful, and in Mr. Hogarth's analysis of Beauty; a new edition of which is much wanted separate from his other works.

The sentiment of Beauty appears to be attached from our cradles to the easy curvatures of lines, and smooth surfaces of visible objects, and to have been derived from the form of the female bosom; as spoken of in Zoonomia, Vol. I, Section XVI, on Instinct.

Sentimental love, as distinguished from the animal passion of that name, with which it is frequently accompanied, consists in the desire or sensation of beholding, embracing, and saluting, a beautiful object.

The characteristic of beauty therefore is that it is the object of love; and though many other objects are in common language called beautiful, yet they are only called so metaphorically, and ought to be termed agreeable. A Grecian temple may give us the pleasurable idea of sublimity; a Gothic temple may give us the pleasurable idea of variety; and a modern house the pleasurable idea of utility; music and poetry may inspire our love by association of ideas; but none of these, except metaphorically, can be termed beautiful; as we have no wish to embrace or salute them.

Our perception of beauty consists in our recognition by the sense of vision of those objects, first which have before inspired our love by the pleasure, which they have afforded to many of our senses: as to our sense of warmth, of touch, of smell, of taste, hunger and thirst; and secondly, which bear any analogy of form to such objects.

When the babe, soon after it is born into this cold world, is applied to its mother's bosom, its sense of perceiving warmth is first agreeably affected; next its sense of smell is delighted with the odour of her milk; then its taste is gratified by the flavour of it, afterwards the appetites of hunger and of thirst afford pleasure by the possession of their objects, and by the subsequent digestion of the aliment; and lastly, the sense of touch is delighted by the softness and smoothness of the milky fountain, the source of such variety of happiness.

XIV.
THE THEORY AND STRUCTURE OF LANGUAGE

> Next to each thought associate sound accords,
> And forms the dulcet symphony of words.
> Canto III, line 365

Ideas consist of synchronous motions or configurations of the extremities of the organs of sense; these when repeated by sensation, volition, or association, are either simple or complex, as they were first excited by irritation; or have afterwards some parts abstracted from them, or some parts added to them. Language consists of words, which are the names or symbols of ideas. Words are therefore properly all of them nouns or names of things.

Little had been done in the investigation of the theory of language from the time of Aristotle to the present aera, till Mr. Horne Tooke, the ingenious and learned author of the Diversions of Purley, explained those undeclined words of all languages, which had puzzled the grammarians, and evinced from their etymology, that they were abbreviations of other modes of expression. Mr. Tooke observes, that the first aim of language was to communicate our thoughts, and the second to do it with dispatch; and hence he divides words into those, which were necessary to express our thoughts, and those which are abbreviations of the former; which he ingeniously styles the wings of Hermes.

For the greater dispatch of conversation many words suggest more than one idea; I shall therefore arrange them according to the number and kinds of ideas, which they suggest; and am induced to do this, as a new distribution of the objects of any science may advance the knowledge of it by developing another analogy of its constituent parts. And in thus endeavouring to analyze the theory of language I mean to speak primarily of the English, and occasionally to add what may occur concerning the structure of the Greek and Latin.

I. *Conjunctions and Prepositions*

The first class of words consists of those, which suggest but one idea, and suffer no change of termination; which have been termed by grammarians Conjunctions and Prepositions; the former of which connect sentences, and the latter words. Both which have been ingeniously explained by Mr. Horne Tooke from their etymology to be abbreviations of other modes of expression.

1. Thus the conjunction *if* and *an*, are shown by Mr. Tooke to be derived from the imperative mood of the verbs to give and to grant; but both of these conjunctions by long use appear to have become the name of a more abstracted idea, than the words give or grant suggest, as they do not now express any ideas of person, or of number, or of time; all which are generally attendant upon the meaning of a verb; and perhaps all the words of this class are the names of ideas much abstracted, which has caused the difficulty of explaining them.

2. The number of Prepositions is very great in the English language, as they are used before the cases of nouns, and the infinitive mood of verbs, instead of the numerous changes of termination of the nouns and verbs of the Greek and Latin; which gives greater simplicity to our language, and greater facility of acquiring it.

The prepositions, as well as the preceding conjunctions, have been well explained by Mr. Horne Tooke; who has developed the etymology of many of them. As the greatest number of the ideas, we receive from external objects, are complex ones, the names of these constitute a great part of language, as the proper names of persons and places; which are complex terms. Now as these complex terms do not always exactly suggest the quantity of combined ideas we mean to express, some of the prepositions are prefixed to them to add or to deduct something, or to limit their general meaning; as a house with a party wall, or a house without a roof. These words are also derived by Mr. Tooke, as abbreviations of the imperative moods of verbs; but which appear now to suggest ideas further abstracted than those generally suggested by verbs, and are all of them properly nouns, or names of ideas.

II. *Nouns Substantive*

The second class of words consists of those, which in their simplest state suggest but one idea, as the word man; but which by two changes of termination in our language suggest one secondary idea of number, as the word men; or another secondary idea of the genitive case, as man's mind, or the mind of man. These words by other changes of termination in the Greek and Latin languages suggest many other secondary ideas, as of gender, as well as of number, and of all the other cases described in their grammars; which in English are expressed by prepositions.

This class of words includes the Nouns Substantive, or names of things, of common grammars, and may be conveniently divided into three kinds. 1. Those which suggest the ideas of things believed to possess hardness and figure, as a house or a horse. 2. Those which suggest the ideas of things, which are not supposed to possess hardness and figure, except metaphorically, as virtue, wisdom; which have therefore been termed abstracted ideas. 3. Those which have been called by metaphysical writers reflex ideas, and mean those of the operations of the mind, as sensation, volition, association.

Another convenient division of these nouns substantive or names of things may be first into general terms, or the names of classes of ideas, as man, quadruped, bird, fish, animal. 2. Into the names of complex ideas, as this house, that dog. 3. Into the names of simple ideas, as whiteness, sweetness.

A third convenient division of the names of things may be into the names of intire things, whether of real or imaginary being; these are the nouns substantive of grammars. 2. Into the names of the qualities or properties of the former; these are the nouns adjective of grammars. 3. The names of more abstracted ideas as the conjunctions and prepositions of grammarians.

These nouns substantive, or names of intire things, suggest but one idea in their simplest form, as in the nominative case singular of grammars. As the word a stag is the name of a single complex idea; but the word stags by a change of termination adds to this a secondary idea of number; and the word stag's, with a comma before the final s, suggests, in English, another secondary idea of something appertaining to the stag, as a stag's horn; which is, however, in our language, as frequently expressed by the preposition *of*, as the horn of a stag.

In the Greek and Latin languages an idea of gender is joined with the names of intire things, as well as of number; but in the English language the nouns, which express inanimate objects, have no genders except metaphorically; and even the sexes of many animals have names so totally different from each other, that they rather give an idea of the individual creature than of the sex, as bull and cow, horse and mare, boar and sow, dog and bitch. This constitutes another circumstance, which renders our language more simple, and more easy to acquire; and at the same time contributes to the poetic excellence of it; as by adding a masculine or feminine pronoun, as he, or she, other nouns substantive are so readily personified.

In the Latin language there are five cases besides the nominative, or original word, and in the Greek four. Whence the original noun substantive by change of its termination suggests a secondary idea either corresponding with the genitive, dative, accusative, vocative, or ablative cases, besides the secondary ideas of number and gender above mentioned. The ideas suggested by these changes of termination, which are termed cases, are explained in the grammars of these languages, and are expressed in ours by prepositions, which are called the signs of those cases.

Thus the word Domini, of the Lord, suggests beside the primary idea a secondary one of something appertaining to it, as templum domini, the temple of the Lord, or the Lord's temple; which in English is either effected by an addition of the letter s, with a comma before it, or by the preposition *of*. This genitive case is said to be expressed in the Hebrew language simply by the locality of the words in succession to each other; which must so far add to the conciseness of that language.

Thus the word Domino, in the dative case, to the Lord, suggests besides the primary idea a secondary one of something being added to the primary one; which is effected in English by the preposition *to*.

The accusative case, or Dominum, besides the primary idea implies something having acted upon the object of that primary idea; as felis edit murem, the cat eats the mouse. This is thus effected in the Greek and Latin by a change of termination of the noun acted upon, but is managed in a more concise way in our language by its situation in the sentence, as it follows the verb. Thus if the mouse in the above sentence was placed before the verb, and the cat after it, in English the sense would be inverted, but not so in Latin; this necessity of generally placing the accusative case after the verb is inconvenient in poetry; though it adds to the conciseness and simplicity of our language, as it saves the intervention of a preposition, or of a change of termination.

The vocative case of the Latin language, or Domine, besides the primary idea suggests a secondary one of appeal, or address; which in our language is either marked by its situation in the sentence, or by the preposition O preceding it. Whence this interjection O conveys the idea of appeal joined to the subsequent noun, and is therefore properly another noun, or name of an idea, preceding the principal one like other prepositions.

The ablative case in the Latin language, as Domino, suggests a secondary idea of something being deducted from or by the primary one. Which is perhaps more distinctly expressed by one of those prepositions in our language; which, as it suggests somewhat concerning the adjoined noun, is properly another noun, or name of an idea, preceding the principal one.

When to these variations of the termination of nouns in the singular number are added those equally numerous of the plural, and the great variety of these terminations correspondent to the three genders, it is evident that the prepositions of our own and other modern languages instead of the changes of termination add to the simplicity of these languages, and to the facility of acquiring them.

Hence in the Latin language, besides the original or primary idea suggested by each noun substantive, or name of an entire thing, there attends an additional idea of number, another of gender, and another suggested by each change of termination, which constitutes the cases; so that in this language four ideas are suggested at the same time by one word; as the primary idea, its gender, number, and case; the latter of which has also four or five varieties. These nouns therefore may properly be termed the abbreviation of sentences; as the conjunctions and prepositions are termed by Mr. Tooke the abbreviation of words; and if the latter are called the wings affixed to the feet of Hermes, the former may be called the wings affixed to his cap.

III. *Adjectives, Articles, Participles, Adverbs*

1. The third class of words consists of those, which in their simplest form suggest two ideas; one of them is an abstracted idea of the quality of an object, but not of the object itself; and the other is an abstracted idea of its appertaining to some other noun called a substantive, or a name of an entire thing.

These words are termed Adjectives, are undeclined in our language in respect to cases, number, or gender; but by three changes of termination they suggest the secondary ideas of greater, greatest, and of less; as the word sweet changes into sweeter, sweetest, and sweetish; which may be termed three degrees of comparison besides the positive meaning of the word; which terminations of *er* and *est* are seldom added to words of more than two syllables; as those degrees are then most frequently denoted by the prepositions more and most.

Adjectives seem originally to have been derived from nouns substantive, of which they express a quality, as a musky rose, a beautiful lady, a stormy day. Some of them are formed from the correspondent substantive by adding the syllable *ly*, or *like*, as a lovely child, a warlike countenance; and in our language it is frequently only necessary to put a hyphen between two nouns substantive for the purpose of converting the former one into an adjective, as an eagle-eye, a Mayday. And many of our adjectives are substantives unchanged, and only known by their situation in a sentence, as a German, or a German gentleman. Adjectives therefore are names of qualities, or parts of things; as substantives are the names of entire things.

In the Latin and Greek languages these adjectives possess a great variety of terminations; which suggest occasionally the ideas of number, gender, and the various cases, agreeing in all these with the substantive, to which they belong; besides the two original or primary ideas of quality, and of their appertaining to some other word, which must be adjoined to make them sense. Insomuch that some of these adjectives, when declined through all their cases, and genders, and numbers, in their positive, comparative, and superlative degrees, enumerate fifty or sixty terminations. All which to one, who wishes to learn these languages, are so many new words, and add much to the difficulty of acquiring them.

Though the English adjectives are undeclined, having neither case, gender, nor number; and with this simplicity of form possess a degree of comparison by the additional termination of ish, more than the generality of Latin or Greek adjectives, yet are they less adapted to poetic measure, as they must accompany their corresponding substantives; from which they are perpetually separated in Greek and Latin poetry.

2. There is a second kind of adjectives, which abound in our language, and in the Greek, but not in the Latin, which are called Articles by the writers of grammar, as the letter *a*, and the word *the*. These, like the adjectives above described, suggest two primary ideas, and suffer no change of termination in our language, and therefore suggest no secondary ideas.

Mr. Locke observes, that languages consist principally of general terms; as it would have been impossible to give a name to every individual object, so as to communicate an idea of it to others; it would be like reciting the name of every individual soldier of an army, instead of using the general term, army. Now the use of the article *a*, and *the* in English, and *o* in Greek, converts general terms into particular ones; this idea of particularity as a quality, or property of a noun, is one of the primary ideas suggested by these articles; and the other is, that of its appertaining to some particular noun substantive, without which it is not intelligible. In both these respects these articles correspond with adjectives; to which may be added, that our article *a* may be expressed by the adjective one or any; and that the Greek article *o* is declined like other adjectives.

The perpetual use of the article, besides its converting general terms into particular ones, contributes much to the force and beauty of our language from another circumstance, that abstracted ideas become so readily personified simply by the omission of it; which perhaps renders the English language better adapted to poetry than any other ancient or modern: the following prosopopoeia from Shakspeare is thus beautiful.

> She let Concealment like a worm i' th' bud
> Feed on her damask cheek.

And the following line, translated from Juvenal by Dr. Johnson, is much superior to the original, owing to the easy personification of Worth and Poverty, and to the consequent conciseness of it.

> Difficile emergunt, quorum virtutibus obstat
> Res angusta domi.

> Slow rises Worth by Poverty depress'd.

3. A third class of adjectives includes what are termed Participles, which are allied to the infinitive moods of verbs, and are formed in our language by the addition only of the syllable *ing* or *ed*; and are of two kinds, active and passive, as loving, loved, from the verb to love. The verbs suggest an idea of the noun, or thing spoken of; and also of its manner of existence, whether at rest, in action, or in being acted upon; as I lie still, or I whip, or I am whipped; and, lastly, another idea of the time of resting, acting, or suffering; but these adjectives called participles, suggest only two primary ideas, one of the noun, or thing spoken of, and another of the mode of existence, but not a third idea of time; and in this respect participles differ from the verbs, from which they originate, or which originated from them, except in their infinitive moods.

Nor do they resemble adjectives only in their suggesting but two primary ideas; but in the Latin and Greek languages they are declined through all the cases, genders, and numbers, like other adjectives; and change their terminations in the degrees of comparison.

In our language the participle passive, joined to the verb to be, for the purpose of adding to it the idea of time, forms the whole of the passive voice; and is frequently used in a similar manner in the Latin language, as I am loved is expressed either by amor, or amatus sum. The construction of the whole passive voice from the verb to be and the participles passive of other verbs, contributes much to the simplicity of our language, and the ease of acquiring it; but renders it less concise than perhaps it might have been by some simple variations of termination, as in the active voice of it.

4. A fourth kind of adjective is called by the grammarians an Adverb; which has generally been formed from the first kind of adjectives, as these were frequently formed from correspondent substantives; or it has been formed from the third kind of adjectives, called participles; and this is effected in both cases by the addition, of the syllable *ly*, as wisely, charmingly.

This kind of adjective suggests two primary ideas, like the adjectives, and participles, from which they are derived; but differ from them in this curious circumstance, that the other adjectives relate to substantives, and are declined like them in the Latin and Greek languages, as a lovely boy, a warlike countenance; but these relate to verbs, and are therefore undeclined, as to act boldly, to suffer patiently.

IV. *Verbs*

The fourth class of words consists of those which are termed Verbs, and which in their simplest state suggest three ideas; first an idea of the noun, or name of the thing spoken of, as a whip. 2. An idea of its mode of existence, whether at rest, or in action, or in being acted upon. 3. An idea of the time of its existence. Thus "the beadle whipped the beggar," in prolix language might be expressed, the beadle with a whip struck in time past the beggar. Which three ideas are suggested by the one word whipped.

Verbs are therefore nouns, or names of intire ideas, with the additional ideas of their mode of existence and of time; but the participles suggest only the noun, and the mode of existence, without any idea of time; as whipping, or whipped. The infinitive moods of verbs correspond in their signification with the participles; as they also suggest only the noun, or name of the thing spoken of, and an idea of its mode of existence, excluding the idea of time; which is expressed by all the other moods and tenses; whence it appears, that the infinitive mood, as well as the participle, is not truly a part of the verb; but as the participle resembles the adjective in its construction; so the infinitive mood may be said to resemble the substantive, and it is often used as a nominative case to another verb.

Thus in the words "a charming lady with a smiling countenance," the participle acts as an adjective; and in the words "to talk well commands attention," the infinitive mood acts as the nominative case of a noun substantive; and their respective significations are also very similar, as whipping, or to whip, mean the existence of a person acting with a whip.

In the Latin language the verb in its simplest form, except the infinitive mood, and the participle, both which we mean to exclude from complete verbs, suggests four primary ideas, as amo, suggests the pronoun I, the noun love, its existence in its active state, and the present time; which verbs in the Greek and Latin undergo an uncounted variation of termination, suggesting so many different ideas in addition to the four primary ones.

We do not mean to assert, that all verbs are literally derived from nouns in any language; because all languages have in process of time undergone such great variation; many nouns having become obsolete or have perished, and new verbs have been imported from foreign languages, or transplanted from ancient ones; but that this has originally been the construction of all verbs, as well as those to whip and to love above mentioned, and innumerable others.

Thus there may appear some difficulty in analyzing from what noun substantive were formed the verbs to stand or to lie; because we have not properly the name of the abstract ideas from which these verbs arose, except we use the same word for the participle and the noun substantive, as standing, lying. But the verbs, to sit, and to walk, are less difficult to trace to their origin; as we have names for the nouns substantive, a seat, and a walk.

But there is another verb of great consequence in all languages, which would appear, in its simplest form in our language to suggest but two primary ideas, as the verb to be, but that it suggests three primary ideas like other verbs maybe understood, if we use the synonymous term to exist instead of *to be*. Thus "I exist" suggests first the abstract idea of existence, not including the mode of existence, whether at rest, or in action, or in suffering; secondly it adds to that abstracted idea of existence its real state, or actual resting, acting, or suffering, existence; and thirdly the idea of the present time: thus the infinitive mood *to be*, and the participle, *being*, suggest both the abstract idea of existence, and the actual state of it, but not the time.

The verb *to be* is also used irregularly to designate the parts of time and actual existence; and is then applied to either the active or passive participles of other verbs, and called an auxiliary verb; while the mode of existence, whether at rest, or in action, or being acted upon, is expressed by the participle, as "I am loving" is nearly the same as "I love," amo; and "I am loved," amatus sum, is nearly the same as amor. This mode of application of the verb *to be* is used in French as well as in English, and in the passive voice of the Latin, and perhaps in many other languages; and is by its perpetual use in conversation rendered irregular in them all, as I am, thou art, he is, would not seem to belong to the infinitive mood *to be*, any more than sum, fui, sunt, fuerunt, appear to belong to esse.

The verb *to have* affords another instance of irregular application; the word means in its regular sense to possess, and then suggests three ideas like the above verb of existence: first the abstracted idea of the thing spoken of, or possession; secondly, the actual existence of possession, and lastly the time, as I have or possess. This verb *to have* like the verb *to be* is also used irregularly to denote parts of past time, and is then joined to the passive participles alone, as I have eaten; or it is accompanied with the passive participle of the verb to be, and then with the active participle of another verb, as I have been eating.

There is another word *will* used in the same irregular manner to denote the parts of future time, which is derived from the verb *to will*; which in its regular use signifies to exert our volition. There are other words used to express other circumstances attending upon verbs, as may, can, shall, all which are probably the remains of verbs otherwise obsolete. Lastly, when we recollect, that in the moods and tenses of verbs one word expresses never less than three ideas in our language, and many more in the Greek and Latin; as besides those three primary ideas the idea of person, and of number, are always expressed in the indicative mood, and other ideas suggested in the other moods, we cannot but admire what excellent abbreviations of language are thus achieved; and when we observe the wonderful intricacy and multiplicity of sounds in those languages, especially in the Greek verbs, which change both the beginning and ending of the original word through three voices, and three numbers, with uncounted variations of dialect; we cannot but admire the simplicity of modern languages compared to these ancient ones; and must finally perceive, that all language consists simply of nouns, or names of ideas, disposed in succession or in combination, all of which are expressed by separate words, or by various terminations of the same word.

Conclusion

The theory of the progressive production of language in the early times of society, and its gradual improvements in the more civilized ones, may be readily induced from the preceding pages. In the commencement of Society the names of the ideas of entire things, which, it was necessary most frequently to communicate, would first be invented, as the names of individual persons, or places, fire, water, this berry, that root; as it was necessary perpetually to announce, whether one or many of such external things existed, it was soon found more convenient to add this idea of number by a change of termination of the word, than by the addition of another word.

As many of these nouns soon became general terms, as bird, beast, fish, animal; it was next convenient to distinguish them when used for an individual, from the same word used as a general term; whence the two articles *a* and *the*, in our language, derive their origin.

Next to these names of the ideas of entire things, the words most perpetually wanted in conversation would probably consist of the names of the ideas of the parts or properties of things; which might be derived from the names of some things, and applied to others which in these respects resembled them; these are termed adjectives, as rosy cheek, manly voice, beastly action; and seem at first to have been formed simply by a change of termination of their correspondent substantives. The comparative degrees of greater and less were found so frequently necessary to be suggested, that a change of termination even in our language for this purpose was produced; and is as frequently used as an additional word, as wiser or more wise.

The expression of general similitude, as well as partial similitude, becomes so frequently used in conversation, that another kind of adjective, called an adverb, was expressed by a change of termination, or addition of the syllable ly or like; and as adjectives of the former kind are applied to substantives, and express a partial similitude, these are applied to verbs and express a general similitude, as to act heroically, to speak boldly, to think freely.

The perpetual chain of causes and effects, which constitute the motions, or changing configurations, of the universe, are so conveniently divided into active and passive, for expressing the exertions or purposes of common life, that it became particularly convenient in all languages to substitute changes of termination, instead of additional nouns, to express, whether the thing spoken of was in a state of acting or of being acted upon. This change of termination betokening action or suffering constitutes the participle, as loving, loved; which, as it expresses a property of bodies, is classed amongst adjectives in the preceding pages.

Besides the perpetual allusions to the active or passive state of things, the comparative times of these motions, or changes, were also perpetually required to be expressed; it was therefore found convenient in all languages to suggest them by changes of terminations in preference to doing it by additional nouns. At the same time the actual or real existence of the thing spoken of was perpetually required, as well as the times of their existence, and the active or passive state of that existence. And as no conversation could be carried on without unceasingly alluding to these circumstances, they became in all languages suggested by changes of termination; which are termed moods and tenses in grammars, and convert the participle above mentioned into a verb; as that participle had originally been formed by adding a termination to a noun, as chaining, and chained, from chain.

The great variety of changes of termination in all languages consists therefore of abbreviations used instead of additional words; and adds much to the conciseness of language, and the quickness with which we are enabled to communicate our ideas; and may be said to add unnumbered wings to every limb of the God of Eloquence.

XV.
ANALYSIS OF ARTICULATE SOUNDS

The tongue, the lips articulate; the throat
With soft vibration modulates the note.
Canto III, line 367

Having explained in the preceding account of the theory of language that it consists solely of nouns, or the names of ideas, disposed in succession or combination; I shall now attempt to investigate the number of the articulate sounds, which constitute those names of ideas by their successions and combinations; and to show by what parts of the organs of speech they are modulated and articulated; whence may be deduced the precise number of letters or symbols necessary to suggest those sounds, and form an alphabet, which may spell with accuracy the words of all languages.

I. *Imperfections of the present Alphabet*

It is much to be lamented, that the alphabet, which has produced and preserved almost all the improvements in other arts and sciences, should have itself received no improvement in modern times; which have added so much elucidation to almost every branch of knowledge, that can meliorate the condition of humanity. Thus in our present alphabets many letters are redundant, others are wanted; some simple articulate sounds have two letters to suggest them; and in other instances two articulate sounds are suggested by one letter. Some of these imperfections in the alphabet of our own language shall be enumerated.

X. Thus the letter x is compounded of ks, or of gz, as in the words excellent, example: eksellent, egzample.

C is sometimes k, at other times s, as in the word access.

G is a single letter in go; and suggests the letters d and the French J in pigeon.

Qu is kw, as quality is kwality.

NG in the words long and in king is a simple sound like the French n, and wants a new character.

SH is a simple sound, and wants a new character.

TH is either sibilant as in thigh; or semivocal as in thee; both of which are simple sounds, and want two new characters.

J French exists in our words confusion, and conclusion, judge, pigeon, and wants a character.

J consonant, in our language, expresses the letters d, and the French j conjoined, as in John, Djon.

CH is either k as in Arch-angel, or is used for a sound compounded of Tsh, as in Children, Tshildren.

GL is dl, as Glove is pronounced by polite people dlove.

CL is tl, as Cloe is pronounced by polite speakers Tloe.

The spelling of our language in respect to the pronunciation is also wonderfully defective, though perhaps less so than that of the French; as the words slaughter and laughter are pronounced totally different, though spelt alike. The word sough, now pronounced suff, was formerly called sow; whence the iron fused and received into a sough acquired the name of sowmetal; and that received into less soughs from the former one obtained the name of pigs of iron or of lead; from the pun on the word sough, into sow and pigs. Our word jealousies contains all the vowels, though three of them only were necessary; nevertheless in the two words abstemiously and facetiously the vowels exist all of them in their usual order, and are pronounced in their most usual manner.

Some of the vowels of our language are diphthongs, and consist of two vocal sounds, or vowels, pronounced in quick succession; these diphthongs are discovered by prolonging the sound, and observing, if the ending of it be different from the beginning; thus the vowel i in in our language, as in the word high, if drawn put ends in the sound of the letter e as used in English; which is expressed by the letter i in most other languages: and the sound of this vowel i begins with ah, and consists therefore of ah and ee. Whilst the diphthong on in our language, as in the word how, begins with ah also and ends in oo, and the vowel u of our language, as in the word use, is likewise a diphthong; which begins with e and ends with oo, as eoo. The French u is also a diphthong compounded of a and oo, as aoo. And many other defects and redundancies in our alphabet will be seen by perusing the subsequent structure of a more perfect one.

II. *Production of Sounds*

By our organ of hearing we perceive the vibrations of the air; which vibrations are performed in more or in less time, which constitutes high or low notes in respect to the gammut; but the tone depends on the kind of instrument which produces them. In speaking of articulate sounds they may be conveniently divided first into clear continued sounds, expressed by the letters called vowels; secondly, Into hissing sounds, expressed by the letters called sibilants; thirdly, Into semivocal sounds, which consist of a mixture of the two former; and, lastly, Into interrupted sounds, represented by the letters properly termed consonants.

The clear continued sounds are produced by the streams of air passing from the lungs in respiration through the larynx; which is furnished with many small muscles, which by their action give a proper tension to the extremity of this tube; and the sounds, I suppose, are produced by the opening and closing of its aperture; something like the trumpet stop of an organ, as may be observed by blowing through the wind-pipe of a dead goose.

These sounds would all be nearly similar except in their being an octave or two higher or lower; but they are modulated again, or acquire various tones, in their passage through the mouth; which thus converts them into eight vowels, as will be explained below.

The hissing sounds are produced by air forcibly pushed through certain passages of the mouth without being previously rendered sonorous by the larynx; and obtain their sibilancy from their slower vibrations, occasioned by the mucous membrane, which lines those apertures or passages, being less tense than that of the larynx. I suppose the stream of air is in both cases frequently interrupted by the closing of the sides or mouth of the passages or aperture; but that this is performed much slower in the production of sibilant sounds, than in the production of clear ones.

The semivocal sounds are produced by the stream of air having received quick vibrations, or clear sound, in passing through the larynx, or in the cavity of the mouth; but apart of it, as the outsides of this sonorous current of air, afterwards receives slower vibrations, or hissing sound, from some other passages of the lips or mouth, through which it then flows. Lastly the stops, or consonants, impede the current of air, whether sonorous or sibilant, for a perceptible time; and probably produce some change of tone in the act of opening and closing their apertures.

There are other clear sounds besides those formed by the larynx; some of them are formed in the mouth, as may be heard previous to the enunciation of the letters b, and d, and ga; or during the pronunciation of the semivocal letters, v. z. j. and others in sounding the liquid letters r and l; these sounds we shall term orisonance. The other clear sounds are formed in the nostrils, as in pronouncing the liquid letters m, n, and ng, these we shall term narisonance.

Thus the clear sounds, except those above mentioned, are formed in the larynx along with the musical height or lowness of note; but receive afterward a variation of tone from the various passages of the mouth: add to these that as the sibilant sounds consist of vibrations slower than those formed by the larynx, so a whistling through the lips consists of vibrations quicker than those formed by the larynx.

As all sound consists in the vibrations of the air, it may not be disagreeable to the reader to attend to the immediate causes of those vibrations. When any sudden impulse is given to an elastic fluid like the air, it acquires a progressive motion of the whole, and a condensation of the constituent particles, which first receive the impulse; on this account the currents of the atmosphere in stormy seasons are never regular, but blow and cease to blow by intervals; as a part of the moving stream is condensed by the projectile force; and the succeeding part, being consequently rarefied, requires some time to recover its density, and to follow the former part: this elasticity of the air is likewise the cause of innumerable eddies in it; which are much more frequent than in streams of water; as when it is impelled against any oblique plane, it results with its elastic force added to its progressive one.

Hence when a vacuum is formed in the atmosphere, the sides of the cavity forcibly rush together both by the general pressure of the superincumbent air, and by the expansion of the elastic particles of it; and thus produce a vibration of the atmosphere to a considerable distance: this occurs, whether this vacuity of air be occasioned by the discharge of cannon, in which the air is displaced by the sudden evolution of heat, which as suddenly vanishes; or whether the vacuity be left by a vibrating string, as it returns from each side of the arc, in which it vibrates; or whether it be left under the lid of the valve in the trumpet stop of an organ, or of a child's play trumpet, which continues perpetually to open and close, when air is blown through it; which is caused by the elasticity of the currents, as it occasions the pausing gusts of wind mentioned above.

Hence when a quick current of air is suddenly broken by any intervening body, a vacuum is produced by the momentum of the proceeding current, between it and the intervening body; as beneath the valve of the trumpet-stop above mentioned; and a vibration is in consequence produced; which with the great facility, which elastic fluids possess of forming eddies, may explain the production of sounds by blowing through a fissure upon a sharp edge in a common organ-pipe or child's whistle; which has always appeared difficult to resolve; for the less vibration an organ-pipe itself possesses, the more agreeable, I am informed, is the tone; as the tone is produced by the vibration of the air in the organ pipe, and not by that of the sides of it; though the latter, when it exists, may alter the tone though, not the note, like the belly of a harpsichord, or violin.

When a stream of air is blown on the edge of the aperture of an organ-pipe about two thirds of it are believed to pass on the outside of this edge, and one third to pass on the inside of it; but this current of air on the inside forms an eddy, whether the bottom of the pipe be closed or not; which eddy returns upwards, and strikes by quick intervals against the original stream of air, as it falls on the edge of the aperture, and forces outwards this current of air with quick repetitions, so as to make more than two thirds of it, and less than two thirds alternately pass on the outside; whence a part of this stream of air, on each side of the edge of the aperture is perpetually stopped by that edge; and thus a vacuum and vibration in consequence, are reciprocally produced on each side of the edge of the aperture.

The quickness or slowness of these vibrations constitute the higher and lower notes of music, but they all of them are propagated to distant places in the same time; as the low notes of a distant ring of bells are heard in equal times with the higher ones: hence in speaking at a distance from the auditors, the clear sounds produced in the larynx by the quick vibrations of its aperture, which form the vowels; the tremulous sounds of the L, R, M, N, NG, which are owing to vibrations of certain apertures of the mouth and nose, and are so slow, that the intervals between them are perceived; the sibilant sounds, which I suppose are occasioned by the air not rushing into a complete vacuum, whence the vibrations produced are defective in velocity; and lastly the very high notes made by the quickest vibrations of the lips in whistling; are all heard in due succession without confusion; as the progressive motions of all sounds I believe travel with equal velocity, notwithstanding the greater or less quickness of their vibrations.

III. *Structure of the Alphabet*

Mute and antesonant Consonants, and nasal Liquids.

P. If the lips be pressed close together and some air be condensed in the mouth behind them, on opening the lips the mute consonant P begins a syllable; if the lips be closed suddenly during the passage of a current of air through them, the air becomes condensed in the mouth behind them, and the mute consonant P terminates a syllable.

B. If in the above situation of the lips a sound is previously produced in the mouth, which may be termed orisonance, the semisonant consonant B is produced, which like the letter P above described may begin or terminate a syllable.

M. In the above situation of the lips, if a sound is produced through the nostrils, which sound is termed narisonance, the nasal letter M is formed; the sound of which may be lengthened in pronunciation like those of the vowels.

T. If the point of the tongue be applied to the forepart of the palate, at the roots of the upper teeth, and some air condensed in the mouth behind, on withdrawing the tongue downwards the mute consonant T is formed; which may begin or terminate a syllable.

D. If the tongue be placed as above described, and a sound be previously produced in the mouth, the semisonant consonant D is formed, which may begin or terminate a syllable.

N. If in the above situation of the tongue and palate a sound be produced through the nostrils, the nasal letter N is formed, the sound of which may be elongated like those of the vowels.

K. If the point of the tongue be retracted, and applied to the middle part of the palate; and some air condensed in the mouth behind; on withdrawing the tongue downwards the mute consonant K is produced, which may begin or terminate a syllable.

Ga. If in the above situation of the tongue and palate a sound be previously produced in the mouth behind, the semisonant consonant G is formed, as pronounced in the word go, and may begin or terminate a syllable.

NG. If in the above situation of the tongue and palate a sound be produced through the nostrils; the nasal letter ng is produced, as in king and throng; which is the french n, the sound of which may be elongated like a vowel; and should have an appropriated character, as thus ᴎ.

Three of these letters, P, T, K, are stops to the stream of vocal air, and are called mutes by grammarians; three, B, D, Ga, are preceded by a little orisonance; and three, M, N, NG, possess continued narisonance, and have been called liquids by grammarians.

Sibilants and Sonisibilants.

W. Of the Germans; if the lips be appressed together, as informing the letter P; and air from the mouth be forced between them; the W sibilant is produced, as pronounced by the Germans, and by some of the inferiour people of London, and ought to have an appropriated character as thus ⋀.

W. If in the above situation of the lips a sound be produced in the mouth, as in the letter B, and the sonorous air be forced between them; the sonisibilant letter W is produced; which is the common W of our language.

F. If the lower lip be appressed to the edges of the upper teeth, and air from the mouth be forced between them, the sibilant letter F is formed.

V. If in the above situation of the lip and teeth a sound be produced in the mouth, and the sonorous air be forced between them, the sonisibilant letter V is formed.

Th. Sibilant. If the point of the tongue be placed between the teeth, and air from the mouth be forced between them, the Th sibilant is produced, as in thigh, and should have a proper character, as Φ.

Th. Sonisibilant. If in the above situation of the tongue and teeth a sound be produced in the mouth, and the sonorous air be forced between them, the sonisibilant Th is formed, as in Thee; and should have an appropriated character as Θ.

S. If the point of the tongue be appressed to the forepart of the palate, as in forming the letter T, and air from the mouth be forced between them, the sibilant letter S is produced.

Z. If in the above situation of the tongue and palate a sound be produced in the mouth, as in the letter D, and the sonorous air be forced between them, the sonisibilant letter Z is formed.

SH. If the point of the tongue be retracted and applied to the middle part of the palate, as in forming the letter K, and air from the mouth be forced between them, the letter Sh is produced, which

is a simple sound and ought to have a single character, thus λ.
J. French. If in the above situation of the tongue and palate a sound be produced in the mouth, as in the letter Ga; and the sonorous air be forced between them; the J consonant of the French is formed; which is a sonisibilant letter, as in the word conclusion, confusion, pigeon; it should be called Je, and should have a different character from the vowel i, with which it has an analogy, as thus V.
H. If the back part of the tongue be appressed to the pendulous curtain of the palate and uvula; and air from behind be forced between them; the sibilant letter H is produced.
Ch. Spanish. If in the above situation of the tongue and palate a sound be produced behind; and the sonorous air be forced between them; the Ch Spanish is formed; which is a sonisibilant letter, the same as the Ch Scotch in the words Buchanan and loch: it is also perhaps the Welsh guttural expressed by their double L as in Lloyd, Lluellen; it is a simple sound, and ought to have a single character as Ⅱ.

The sibilant and sonisibilant letters may be elongated in pronunciation like the vowels; the sibilancy is probably occasioned by the vibrations of the air being slower than those of the lowest musical notes. I have preferred the word sonisibilants to the word semivocal sibilants; as the sounds of these sonisibilants are formedl in different apertures of the mouth, and not in the larynx like the vowels.

Orisonant Liquids.

R. If the point of the tongue be appressed to the forepart of the palate, as in forming the letters T, D, N, S, Z, and air be pushed between them so as to produce continued sound, the letter R is formed.
L. If the retracted tongue be appressed to the middle of the palate, as in forming the letters K, Ga, NG, Sh, J French, and air be pushed over its edges so as to produce continued sound, the letter L is formed.

The nasal letters m, n, and ng, are clear tremulous sounds like R and L, and have all of them been called liquids by grammarians. Besides the R and L, above described, there is another orisonant sound produced by the lips in whistling; which is not used in this country as a part of language, and has therefore obtained no character, but is analogous to the R and L; it is also possible, that another orisonant letter may be formed by the back part of the tongue and back part of the palate, as in pronouncing H and Ch, which may perhaps be the Welch Ll in Lloyd, Lluellin.

Four pairs of Vowels

A pronounced like au, as in the word call. If the aperture, made by approximating the back part of the tongue to the uvula and pendulous curtain of the palate, as in forming the sibilant letter H, and the sonisibilant letter Ch Spanish, be enlarged just so much as to prevent sibilancy; and a continued sound produced by the larynx be modulated in passing through it; the letter A is formed, as in ball, wall, which is sounded like aw in the word awkward; and is the most usual sound of the letter A in foreign languages; and to distinguish it from the succeeding A might be called A micron; as the aperture of the fauces, where it is produced, is less than in the next A.

A pronounced like ah, as in the word hazard. If the aperture of the fauces above described, between the back part of the tongue and the back part of the palate, be enlarged as much as convenient, and a continued sound, produced in the larynx, be modulated in passing through it; the letter A is formed, as in animal, army, and ought to have an appropriated character in our language, as thus ∀. As this letter A is formed by a larger aperture than the former one, it may be called A mega.

A pronounced as in the words cake, ale. If the retracted tongue by approximation to the middle part of the palate, as in forming the letters R, Ga, NG, Sh, J French, L, leaves an aperture just so large as to prevent sibilancy, and sonorous air from the larynx be modulated in passing through it; the letter A is produced, as pronounced in the words whale, sale, and ought to have an appropriated character in our language, as thus ɔ; this is expressed by the letter E in some modern languages, and might be termed E micron; as it is formed by a less aperture of the mouth than the succeeding E.

E pronounced like the vowel a, when short, as in the words emblem, dwelling. If the aperture above described between the retracted tongue and the middle of the palate be enlarged as much as convenient, and sonorous air from the larynx be modulated in passing through it, the letter E is formed, as in the words egg, herring; and as it is pronounced in most foreign languages, and might be called E mega to distinguish it from the preceding E.

I pronounced like e in keel. If the point of the tongue by approximation to the forepart of the palate, as in forming the letters T, D, N, S, Z, R, leaves an aperture just so large as to prevent sibilancy, and sonorous air from the larynx be modulated in passing through it; the vowel I is produced, which is in our language generally represented by e when long, as in the word keel; and by i when, short, as in the word it, which is the sound of this letter in most foreign languages; and may be called E micron to distinguish it from the succeeding E or Y.

Y, when it begins a word, as in youth. If the aperture above described between the point of the tongue, and the forepart of the palate be enlarged as much as convenient, and sonorous air from the larynx be modulated in passing through it, the letter Y is formed; which, when it begins a word, has been called Y consonant by some, and by others has been thought only a quick pronunciation of our e, or the i of foreign languages; as in the word year, yellow; and may be termed E mega, as it is formed by a larger aperture than the preceding e or i.

O pronounced like oo, as in the word fool. If the lips by approximation to each other, as in forming the letters P, B, M, W sibilant, W sonisibilant, leave an aperture just so wide as to prevent sibilancy; and sonorous air from the larynx be modulated in passing through it; the letter O is formed, as in the words cool, school, and ought to have an appropriated character as thus ∞, and may be termed o micron to distinguish it from the succeeding o.

O pronounced as in the word cold. If the aperture above described between the approximated lips be enlarged as much as convenient; and sonorous air from the larynx be modulated in passing through it, the letter o is formed, as in sole, coal, which may be termed o mega, as it is formed in a larger aperture than the preceding one.

Conclusion

The alphabet appears from this analysis of it to consist of thirty-one letters, which spell all European languages.

Three mute consonants, P, T, K.
Three antesonant consonants, B, D, Ga.
Three narisonant liquids, M, N, NG.
Six sibilants, W German, F, Th, S, Sh, H.
Six sonisibilants, W, V, Th, Z, J French, Ch Spanish.
Two orisonant liquids, R, L.
Eight vowels, Aw, ah, a, e, i, y, oo, o.

To these thirty-one characters might perhaps be added one for the Welsh L, and another for whistling with the lips; and it is possible, that some savage nations, whose languages are said to abound with gutturals, may pronounce a mute consonant, as well as an antesonant one, and perhaps another narisonant letter, by appressing the back part of the tongue to the back part of the palate, as in pronouncing the H, and Ch Spanish.

The philosophical reader will perceive that these thirty-one sounds might be expressed by fewer characters referring to the manner of their production. As suppose one character was to express the antesonance of B, D, Ga; another the orisonance of R, L; another the sibilance of W, S, Sh, H; another the sonisibilance of W, Z, J French, Ch Spanish; another to express the more open vowels; another the less open vowels; for which the word micron is here used, and for which the word mega is here used.

Then the following characters only might be necessary to express them all; P alone, or with antesonance B; with narisonance M; with sibilance W German; with sonisibilance W; with vocality, termed micron OO; with vocality, termed mega O.

T alone, or with the above characters added to it, would in the same manner suggest D, N, S, Z, EE, Y, and R with a mark for orisonance.

K alone, or with the additional characters, would suggest Ga, NG, Sh, J French, A, E, and L, with a mark for orisonance.

F alone, or with a mark for sonisibilance, V.

Th alone, or with a mark for sonisibilance, Th.

H alone, or with a mark for sonisibilance, Ch Spanish, and with a mark for less open vocality, aw, with another for more open vocality ah.

Whence it appears that six single characters, for the letters P, T, K, F, Th, H, with seven additional marks joined to them for antesonance, narisonance, orisonance, sibilance, sonisibilance, less open vocality, and more open vocality; being in all but thirteen characters, may spell all the European languages.

I have found more difficulty in analyzing the vowels than the other letters; as the apertures, through which they are modulated, do not close; and it was therefore less easy to ascertain exactly, in what part of the mouth they were modulated; but recollecting that those parts of the mouth must be more ready to use for the purpose of forming the vowels, which were in the habit of being exerted in forming the other letters; I rolled up some tin foil into cylinders about the size of my finger; and speaking the vowels separately through them, found by the impressions made on them, in what part of the mouth each of the vowels was formed with somewhat greater accuracy, but not so as perfectly to satisfy myself.

The parts of the mouth appeared to me to be those in which the letters P, I, K, and H, are produced; as those, where the letters F and Th are formed, do not suit the production of mute or antesonant consonants; as the interstices of the teeth would occasion some sibilance; and these apertures are not adapted to the formation of vowels on the same account.

The two first vowels aw and ah being modulated in the back part of the mouth, it is necessary to open wide the lips and other passages of the mouth in pronouncing them; that those passages may not again alter their tone; and that more so in pronouncing ah, than aw; as the aperture of the fauces is opened wider, where it is formed, and from the greater or less size of these apertures used in forming the vowels by different persons, the tone of all of them may be somewhat altered as spoken by different orators.

I have treated with greater confidence on the formation of articulate sounds, as I many years ago gave considerable attention to this subject for the purpose of improving shorthand; at that time I contrived a wooden mouth with lips of soft leather, and with a valve over the back part of it for nostrils, both which could be quickly opened or closed by the pressure of the fingers, the vocality was given by a silk ribbon about an inch long and a quarter of an inch wide stretched between two bits of smooth wood a little hollowed; so that when a gentle current of air from bellows was blown on the edge of the ribbon, it gave an agreeable tone, as it vibrated between the wooden sides, much like a human voice. This head pronounced the p, b, m, and the vowel a, with so great nicety as to deceive all who heard it unseen, when it pronounced the words mama, papa, map, and pam; and had a most plaintive tone, when the lips were gradually closed. My other occupations prevented me from proceeding in the further construction of this machine; which might have required but thirteen movements, as shown in the above analysis, unless some variety of musical note was to be added to the vocality produced in the larynx; all of which movements might communicate with the keys of a harpsichord or forte piano, and perform the song as well as the accompaniment; or which if built in a gigantic form, might speak so loud as to command an army or instruct a crowd.

I conclude this with an agreeable hope, that now war is ceased, the active and ingenious of all nations will attend again to those sciences, which better the condition of human nature; and that the alphabet will undergo a perfect reformation, which may indeed make it more difficult to trace the etymologies of words, but will much facilitate the acquisition of modern languages; which as science improves and becomes more generally diffused, will gradually become more distinct and accurate than the ancient ones; as metaphors will cease to be necessary in conversation, and only be used as the ornaments of poetry.

THE END

TIMAIOS PRESS

We publish thought-provoking and classic literature for those who are interested in the history of science, philosophy and ideas—books for the general public, as well as students and teachers.

Books by and about:

Epicure — Lucretius — Atomism — Francis Bacon — H.P. Lovecraft — Camille Flammarion — Diogenes Laërtius — Emanuel Swedenborg — Erasmus Darwin — E.T.A. Hoffmann — Plato — Andrew Crosse — And others.

www.timaiospress.com